D0472275

The Factory-Crafted House

The Factory-Crafted House

New Visions of Affordable Home Design

by **Paul S. Sedan**

The
Globe
Pequot
Press

Old Saybrook, Connecticut

Copyright © 1992 by Paul S. Sedan

All rights reserved. No part of this book may be reproduced or transmitted in any form by any means, electronic or mechanical, including photocopying and recording, or by any information storage and retrieval system, except as may be expressly permitted by the 1976 Copyright Act or by the publisher. Requests for permission should be made in writing to The Globe Pequot Press, P.O. Box 833, Old Saybrook, CT 06475.

Photo/Art Credits: Pp. 2, 3: diagrams reprinted with permission from *Automated Builder* magazine, Carpinteria, CA; pp. 5, 14, 31: courtesy The Manufactured Housing Institute; pp. 6 (top), 89: courtesy Alouette Homes, Quebec, Canada; pp. 6 (bottom), 68: courtesy Deltec Homes, Ashville, NC; p. 7 (top): courtesy Real Log Homes, Hartland, VT; p. 7 (bottom): courtesy Cathedralite Domes; pp. 11, 44, 80: courtesy Acorn Structures, Inc., Concord, MA; p. 13: courtesy Wausau Homes, Inc.; pp. 17, 73: courtesy Cardinal Homes, Inc., Wylliesburg, VA; pp. 20, 21: copyright © 1992 Nationwide Homes, Inc.; pp. 39, 42: A.B. Clark, courtesy Miron Truss and Component Corporation, Panelized Home Division, 54 Railroad Avenue, Albany, NY 12205; pp. 45, 102: courtesy Lindal Cedar Homes, Seattle, WA; p. 49: courtesy Alta Log Homes; p. 51: diagram courtesy *Log Home Living* magazine; pp. 52, 54: courtesy Timber Log Homes, Inc., Colchester, CT; pp. 59, 61, 62, 67: courtesy Timberline Geodesics, Berkeley, CA; p. 65: courtesy Maggie Dustin, Classic Post and Beam Homes; p. 77: courtesy Douglas E. Cutler, A.I.A.; p. 83: courtesy New England Log Homes, Inc., Hamden, CT; p. 85: copyright © 1991 Fred Forbes Photography; p. 91: courtesy Unibilt Industries, Inc., Vandalia, OH; p. 97: courtesy Oregon Dome, Inc. All other photographs by the author.

Library of Congress Cataloging-in-Publication Data

Sedan, Paul S.
   The factory-crafted house / by Paul S. Sedan. — 1st ed.
        p.   cm.
     Includes bibliographical references and index.
     ISBN 1-56440-061-1
     1. Prefabricated houses—Design and construction—Amateurs' manuals.   I. Title.
   TH4819.P7S43   1992
   721'.04497—dc20   92-14433
                                                                                          CIP

Front Cover: Acorn Country House Family Model 3300 (courtesy of Acorn Structures, Inc., Concord, MA)
Back Cover: Panelized home photo courtesy Deltec Homes, Ashville, NC; Log home photo courtesy Timber Log Homes, Inc., Colchester, CT; Mobile Home photo courtesy Guerdon Industries, Inc., Lake Oswego, OR; Dome photo courtesy Timberline Geodesics, Berkeley, CA; Modular home photo courtesy Alouette Homes, Quebec, Canada.

Manufactured in the United States of America
First Edition/First Printing

Contents

Acknowledgments

I'd like to thank the following individuals for their help in bringing this book to fruition: Sally McMillan, my agent, for her early encouragement and help; Betsy Amster, my preliminary editor, for her insistence that I keep it simple; Alex Grinnell, of Steven Winter Associates, for helping me to understand more about the industry; Paul Kando, who heads up the Center for the House, for defining the term "factory-crafted"; Mace Lewis, my main man at Globe Pequot for his friendly editorial aid; and all those individuals in the factory-crafted world who gave of their time over the phone or in person.

A special thanks to my dear wife whose full-time work and support made it possible for me to write this book. I am indeed poor without her.

And to carpenters everywhere, in the factory or at the job site, who are in love with the idea of building and the unforgettable smell of fresh-sawn wood.

Introduction

I like the dreams of the future better than the history of the past.
—Thomas Jefferson in a letter to John Adams, 1816

No one has to be told that new homes today are expensive. In fact, depending on the part of the country where you live, you may have found that home ownership is out of your reach.

It is the purpose of this book to help change that by introducing you to factory-crafted homes and the advantages they offer. Right now, factory-crafted homes are available that cost less than their site-built counterparts. Furthermore, factory-crafted homes are generally of a higher quality because they are made under factory-controlled conditions using expensive machinery that works to tight tolerances. And they can be assembled on a lot in far less time than a site-built house because they require less on-site labor.

If you find the idea of a factory-crafted home off-putting, consider this: Nearly every home being built today already uses factory-made components such as doors, windows, stair systems, roof trusses, and decorative moldings in its construction. Even custom builders use these components in their work, relying on more expensive materials or fancier floor plans only to provide that unique "touch."

This book uses the term *factory-crafted* when referring to houses or house components that are built off-site. The term was coined, I believe, by Paul Kando of the Center for the House, in Washington, D.C., and it includes modular, mobile, panelized, and precut homes and components (see page 5 for a more detailed definition of these terms).

It's interesting that when it comes to houses we feel obliged to make these distinctions. We don't refer to a car as "factory-made" because we assume that all cars—with the exception, maybe, of a Ferrari or a Rolls Royce—are made by mass-production techniques. Nor do we refer to factory-made clothing or refrigerators. Yet when it comes to housing, for some reason we feel compelled to cling to the notion of a craft-produced product. Perhaps we want to feel that our home is unique rather than just one among many. Or maybe in our high-tech age we feel the need to cherish the low-tech ideal of an artistic product. Or perhaps we're saying that since we have to deal with machines all day in our work we don't want to come home to another machine at night. Or maybe as we advance technologically we unconsciously want to retreat domestically. (As John Naisbitt says in his book *Megatrends*, we like to combine high-tech with high-touch.)

But as my Uncle John used to say, "regardless of the fact and to the contrary notwithstanding," the housing industry is facing problems that demand solutions. New homes are expensive. Collectively, our income growth is slowing down. And the available labor force—both skilled and unskilled—in the construction industry is shrinking.

As heirs of the industrial revolution, we are used to consumer goods being made cheaper and better by methods of mass-production. That houses should be subject to the same principles seems natural and normal. The time of the factory-crafted home is here. I hope this book helps you see this and makes your dream of home ownership come true.

Due to the variability of local conditions, materials, skills, and job sites, neither The Globe Pequot Press nor the author assumes responsibility for any personal injury, property damage, or other loss of any sort suffered from any actions taken based on, or inspired by, information or advice presented in this book. Mention of a company name does not constitute endorsement. Compliance with local codes and legal requirements is the responsibility of the reader.

1

Why Factory-Crafted?

The last time I got a haircut I was talking to my barber about modular homes.

"Well, I just don't think that I'll ever be interested in one," she said.

"Why not?" I asked. "Do you know anything about them?"

"No," she replied, "but aren't they like a mobile home?"

"Look," I said. "If you were buying a car, which would you rather do? Buy it in parts from the manufacturer and then hire lots of people to put it together for you or buy the complete car from a dealer?"

"Buy it from a dealer, of course!" she answered.

"Well then," I said, "why wouldn't you rather do the same thing with a house? Why buy plans, hire subcontractors, have to put up with delays, bad weather, often poor workmanship, and crazy schedules, when you can have the complete house delivered to your site and put in place in just one day at less cost!"

Now maybe this is an oversimplification. But it does show the pluses of a factory-crafted house. Factory-builders can shorten the time it takes to get a house from the drawing board to your lot. They can significantly cut waste. They can use more sophisticated tools in the assembly operations (which results in a better finished house). And more often than not, they can reduce the cost of the end product—your new home.

Houses from Factories

"Take a skilled carpenter earning $30 an hour at the job site," says Don Carlson, editor and publisher of *Automated Builder* magazine. "Say he has to cut a 2-by-4. The biggest piece of equipment he's got is a hand-held power saw that might cost $60 to $70.

"Compare that with the worker at a component plant. He works on a saw that costs

$52,000, has four blades on one side, three on another, and lumber goes through it at a rate of sixty pieces a minute. This saw cuts within a thousandth of an inch accuracy.

"Whom would you prefer to have cut your lumber?" Actually, the idea for factory-crafted housing is not as new as you might think. In the early 1900s the inventor Thomas A. Edison filed a patent to build concrete houses using cast-iron forms as molds. By 1909 the first such house was completed in Union, New Jersey. It featured 24-inch-thick walls for superb insulation against both heat and cold and cost only $1,600 when finished. Although Edison's model was only a prototype experiment and never met with much success in his day, it paved the way for today's factory-crafted homes.

Today panelized and modular building manufacturers in the United States offer quality materials and design innovations that were unheard of in Edison's time. CAD (computer-aided design) systems are enabling some factory-crafters to offer a wide selection of custom-designed homes. High-tech assembly operations ensure all building materials are of a high quality and consequently that worker delays at the site are limited. Even mobile homes—which today are really not mobile—use up-to-date assembly operations to provide an affordable alternative to costly site-built houses.

The move toward machine made from handmade doesn't have to seem threatening. Today's site-built house already uses an enormous amount of factory-crafted parts in its construction. Bricks, for example, are made by machine. And so are wire nails, wooden moldings,

This diagram demonstrates the simplicity of assembling a modular home compared to a traditional, site-built house, like the one featured in the diagram on page 3.

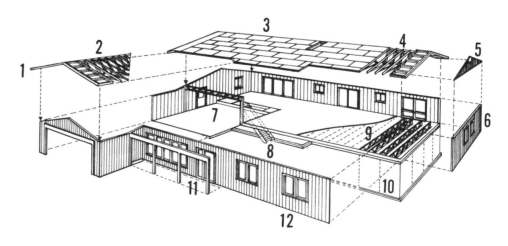

Component Framing Assemblies

1. Ladder Rake overhang assembly
2. Valley Roof trusses
3. Roof Sheathing
4. Engineered roof trusses
5. Gale end
6. Wall panels
7. Garage door header truss
8. Pre-assembled stairs
9. Engineered floor trusses
10. Wood foundation panels
11. Pre-hung doors
12. Pre-hung windows

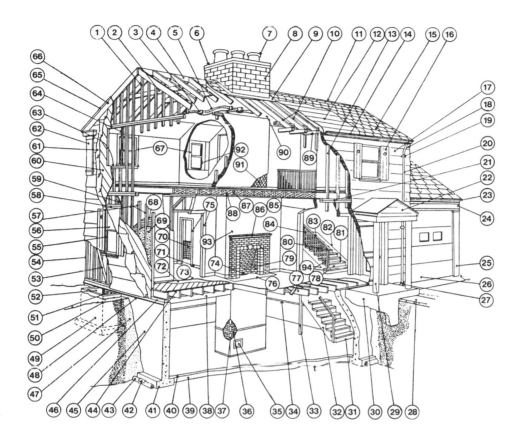

Stick Framing and House Parts

1 – Gable stud
2 – Collar beam
3 – Ceiling joist
4 – Ridge board
5 – Insulation
6 – Chimney cap
7 – Chimney pots
8 – Chimney
9 – Chimney flashing
10 – Rafters
11 – Ridge
12 – Roof boards
13 – Stud
14 – Eave trough or gutter
15 – Roofing
16 – Blind or shutter
17 – Bevel siding
18 – Downspout or leader strap
20 – Downspout leader or Conductor
21 – Double plate
22 – Entrance canopy
23 – Garage cornice
24 – Frieze

25 – Door jamb
26 – Garage door
27 – Downspout or leader shoe
28 – Sidewalk
29 – Entrance post
30 – Entrance platform
31 – Basement stair riser
32 – Stair stringer
33 – Girder post
34 – Chair rail
35 – Cleanout door
36 – Furring strips
37 – Corner stud
38 – Girder
39 – Cinder or gravel fill
40 – Concrete basement floor
41 – Footing for foundation wall
42 – Paper strip
43 – Foundation drain tile
44 – Diagonal subflooring
45 – Foundation wall
46 – Sill

47 – Backfill
48 – Termite shield
49 – Areaway wall
50 – Grade line
51 – Basement sash
52 – Areaway
53 – Corner brace
54 – Corner studs
55 – Window frame
56 – Window light
57 – Wall studs
58 – Header
59 – Window cripple
60 – Wall sheathing
61 – Building paper
62 – Plaster
63 – Rough header
64 – Window stud
65 – Cornice moulding
66 – Frieze or barge board
67 – Window casing
68 – Lath
69 – Insulation
70 – Wainscoting
71 – Baseboard

72 – Building paper
73 – Finish floor
74 – Ash dump
75 – Door trim
76 – Fireplace hearth
77 – Floor joists
78 – Stair riser
79 – Fire brick
80 – Newel cap
81 – Stair tread
82 – Finish stringer
83 – Stair rail
84 – Balusters
85 – Plaster Arch
86 – Mantel
87 – Floor joists
88 – Bridging
89 – Lookout
90 – Attic space
91 – Metal lath
92 – Window sash
93 – Chimney breast
94 – Newel

balustrades, doors and windows, railings, staircases, floor coverings, shingles, plumbing fixtures, and plywood sheathing. Even the fancy pilasters and pediments you see on ersatz colonial or Georgian homes are really factory-made replicas of traditional, hand-carved designs. If you want a truly hand-crafted house, you have to do it yourself or have a ton of money.

This trend toward factory-crafted components makes sense. Machine-made parts are easier and faster to make, so they cost less money. In addition, there are too few craftsmen around today who can still do the work.

Using machines instead of people to make things is commonplace. Mass production is what makes goods affordable. It's the means by which raw materials are turned into such diverse products as toothbrushes, glasses, automobiles, clothing, furniture, telephones, and books.

Housing *components* have been machine made for some time. Even the humble nail is a small example of the way in which the machine age began to revolutionize home building. Until nails were mass-produced, most builders relied on mortise-and-tenon joints and wooden pegs to join pieces of wood. Needless to say, building a house this way took a lot of time.

Changing the System

Where factory-crafting truly promises to shine is in changing the *process* by which most homes today are built.

Since at least the Middle Ages, when large cathedrals were constructed, the typical building team has included a master builder/general contractor who employed a number of tradesmen/subcontractors on an *ad hoc* basis. On a project-by-project basis this type of organization is OK. But when you try to apply it to building a million new homes, it just doesn't work well.

Like all entrepreneurs, subcontractors want to order their activities to make the most profit. This means that if you're a large builder and use the same subcontractors all the time, you can probably count on them to be where you need them, when you need them to be there. But if you're a small builder or an owner, trying to get that drywall crew back to do some sanding or finishing may be next to impossible (at least at *your* convenience). Subs come when it suits *their* schedule, not yours.

Which brings us to the factory-crafted home. In a factory, all workers share a common employer and common projects. They work at independent tasks but do so in a controlled environment. In such an environment all products (houses) can be built to exact specifications and tight tolerances. Homeowners get a product that is completed on time, within budget. And they also get a finished product that will be of higher quality than the one that's site-built.

The Different Types of Factory-Crafted Homes

According to Steven Winter, an architect whose firm specializes in providing design and engineering services to home manufacturers, the term *factory-crafted home* means that "at least 30 to 40 percent of the dwelling unit is prefabricated into completed building components requiring no on-site processes other than connection."

Mobile homes provide the most advanced example of this definition. They leave the factory so complete that they can be wheeled into place and hooked up to existing sewer, water, and electrical connections in a matter of hours.

Today's mobile homes have come a long way from their early predecessors.

Modular homes also leave the factory almost 95 percent complete. They only require a crane to set them on a permanent foundation so they can then be "stitched up"—have minor connections made between modules—and made ready for occupancy.

Panelized homes are the least sophisticated in terms of being preassembled but the most sophisticated in terms of design flexibility. They typically leave the factory as a series of open or closed wall panels of varying sizes. The panels are then connected at the job site to form a "shell," which is then finished like a traditional site-built home. Both *log homes* and *dome homes* are specialized examples of the panelized concept.

Later in this book we will be discussing each type of factory-crafted home in more detail. For now, let us turn our attention to some of the myths surrounding factory-crafted homes.

Modular homes are no longer limited to single-floor ranches or simple two-bedroom capes.

Panelized homes offer the greatest flexibility in the factory-crafted world.

A factory-crafted log home has all the advantages of a handcrafted log home at a savings of up to $20 per square foot.

Even using traditional materials such as shakes and shingles, domes present a distinctive look.

Looking Beyond the Myths About Factory-Crafted Houses

Here are some myths about factory-crafted homes.

Myth: Factory-crafted homes are low-quality and look as though they've been stamped out with a cookie cutter.

Fact: With sophisticated factory machinery, computer controls, better cutting and measuring tools, and more diligent inspection procedures, factory-made components fit together better than parts made by hand on the job site. Designers are just beginning to explore the possibilities of working with factory-built components. Today a factory-crafted home can be a contemporary, colonial, ranch, or cape on the outside and have balconies and cathedral ceilings on the inside.

Myth: Factory-crafted homes don't increase in value the same way site-built homes do.

Fact: When architect Alex Grinnell designed Ellenridge, a townhouse modular development in Ellenville, New York, the units were built to sell for $200,000 each. Today they fetch upward of $350,000 apiece—and no one would ever guess that they are modular.

Myth: Some of the savings realized in factory-crafting are reduced by the shipping costs.

Fact: Lumber and other building components used in stick-building—that is, site-building—have to be shipped, too. And generally they're shipped to a lumberyard, which means paying another middleman. Factory-crafted components are shipped directly to the job site. No middleman. And these components are engineered to stand being moved over the road. This means they are produced to higher standards of construction.

Myth: A good framing crew can stick-build a house just as fast and for less money than it costs to buy and erect a factory-crafted house.

Fact: When the variables of weather, different levels of workmanship, and fluctuating quality of materials from local lumberyards are considered, there is no comparison between a site-built home and one with factory-crafted components. Even if all things were equal, there's no way 20 linear feet of wall can be stick-built in the same amount of time 20 linear feet of panelized wall can be erected. Add to these factors the lack of waste from not having to cut materials on-site and the environmental impact such cutting would have on the local landfill, and you can see that assembling a factory-crafted home is a much more effective method of construction than site-building.

In short, factory-crafted homes are:

- Less expensive
- Faster to build
- Higher quality
- Better engineered
- Flexible in design
- State-of-the-art
- Environmentally friendly

Now that you've been introduced to some of the basic concepts of factory-crafting, read on and find out the details. There could be a factory-crafted home in your future.

2

Selecting a
Factory-Crafted House

Acorn Structures uses panelized components for high-end luxury homes such as this custom Sun-Cape.

If you're in the market for a new home, you've probably considered three options: (1) buying a used home, (2) having one custom-built, or (3) shopping the local subdivisions for products put up by local builders.

If you consider factory-crafted housing, however, a new world of choice opens up for you. In fact, you'll find yourself almost like a kid in a candy store. Your biggest problem will be narrowing your choice.

The five chapters that follow describe the different kinds of factory-crafted homes: modular, mobile, panelized, log, and dome. Each chapter begins with a description of the building concept and is followed by guidelines you can use in making your decision.

To some extent, your budget and the use to which you plan to put your house will determine the type of factory-crafted home you choose:

STARTER HOMES. If you're building your first home and you're watching your money, you may want to look at the following types of homes:

    Mobile
    Modular

MOVE-UP HOMES. If you're already a homeowner and are shopping for a new home, note these categories:

    Modular
    Panelized
    Log

VACATION HOMES. If you're in the market for a second home, almost any of the following could meet your needs:

    Mobile   Log
    Modular   Dome
    Panelized

UNIQUE HOMES. For that special "dream house" take a look at these kinds of homes:

    Panelized
    Log
    Dome

RETIREMENT HOMES. If the kids are gone, your bills are paid, and you want to downsize your style of living, check out these factory-crafted houses:

    Mobile
    Modular

The following table shows just how much time you can save by choosing a factory-crafted home over a site-built structure.

Building Type	Time Needed To Complete House (in weeks)
	5　　10　　15　　20　　25
Site-built	████████████████████████████████
Modular	█
Mobile	
Panelized	████████████████████
Log	███████████████
Dome	████

If we assume an eight-hour workday and a five-day week, we have forty work hours per week. This means a typical modular house, which is brought to the site 95 percent complete, can be finished in about one week, or forty work hours. A typical site-built house requires thirty weeks or 1,200 work hours.

Here's another way of comparing the on-site construction time the two methods take. By the time a person walking at 3 miles per hour walked from New York City to Philadelphia (120 miles), his modular home would be totally complete. On the other hand the same

Once the modules have been joined together, it is impossible to tell the interior of a modular home from that of a site-built home.

person would have to walk all the way from New York City to New Orleans (about 1,300 miles) before his site-built house would be complete.

Since time is money there's no question which home will cost less.

Becoming an Informed Consumer

You will normally follow the same procedure in deciding which factory-crafted home is right for you as you would in buying any other consumer product. But since the price tag is so high (homes are generally the most expensive purchase we make), there are some basic steps you would be wise to take in your decision making.

1. Kick the Tires. Ask if there is a builder/dealer for this company in your area so you can see a model home. If not, where is the closest model?

2. Compare Apples with Apples if Possible. Most factory-crafters are very competitive in their pricing. In comparing costs, you'll want to know what kinds of materials each uses, manner of construction, charges for plans, cost of shipping to your job site, and what is included in the kit or package. This last item is very important. It's hard to compare "apples with apples" unless you know exactly what you're getting for your money. For modular and mobile homes, compare specifications. With other factory-crafted products you will want to find out if you're getting just the basic shell or all that you need to make your house weather tight. Is the roof system complete? Does the package include all

Well-planned developments are a far cry from the old "trailer park" image that the public continues to view with disdain.

doors and windows, or will you have to buy them locally? Nail down all the details so you'll have a good basis of comparison. Companies and their representatives are used to these questions and will be helpful in getting you answers.

3. Sizes and Styles. Does the factory-crafter offer plans in the size and style you want? Most companies offer a wide variety of house plans and styles, from traditional to contemporary to vacation homes. You should be able to find something to suit you. If not, don't give up. A lot of companies are set up to build from custom plans. But you should realize that custom-made houses will cost more than the standard ones.

4. Restrictions. Will you be able to build a factory-crafted home in your area? With the exception of mobile homes, most communities welcome factory-crafted products, and most banks provide temporary (construction) and permanent (mortgage) financing. In addition, manufacturers are very conscious of the regulatory process and have designed and engineered their homes to meet or exceed most state and local building codes.

5. Warranties. Some of the other questions you will want to ask include those about warranties. Most factory-crafters offer limited warranties that cover structural integrity and materials. Check and see what the warranty covers. In addition, most manufacturers' networks of builder/dealers offer at least a one-year warranty on all workmanship. Some are part of the national Home Owners Warranty Program, a ten-year limited warranty/insurance program.

6. References. Most factory-crafters are more than willing to provide you with a list of people in your area who have bought their product and can give you a report on their experiences with it. Call them up and see if they're happy with their new home. Find out if they've had any problems and, if so, what kind.

You really can't go wrong with a factory-crafted home. In most cases it will cost no more (and in many cases it will cost a great deal less) than a comparable site-built house. And today you can get just about any size and style your pocketbook can afford.

3

Modular Homes

Traditional styles are still available in modular homes as exemplified by this four-bedroom cape.

When today's modular house leaves the factory, it's 95 percent complete. Virtually all the plumbing and electrical systems have been installed, kitchen cabinets have been hung, floor coverings have been put down, and interior and exterior surfaces have been painted or papered. When the house reaches the job site, it has only to be installed on its foundation and hooked up to the water, sewer, and electrical supply lines. When this happens, the house is nearly finished, except for some "dealer prep" work (like finishing the areas where the boxes join together) or any additional work the homeowner may have requested such as a porch, deck, garage, or driveway.

A lot of people think that a modular home is like a mobile home. It's not. Mobile homes are built to a uniform national code. Modular homes are more the factory-crafted equivalents of a site-built home—only they go up in a fraction of the time. And since modulars use the same building methods and materials found in their site-built counterparts, it's nearly impossible to tell the difference.

From a builder's point of view, modular homes have cured some of the headaches of the building business. They have virtually eliminated the need to deal with dozens of sub-contractors. They almost totally eliminate waste and debris at the job site. Adverse weather conditions are scarcely a factor at all. And most building code inspections have already taken place at the factory.

From the buyer's point of view, modularity means a better-quality house in less time for less money. It's clearly a win-win situation.

Historically, the concept is nothing new. The word *modular* has to do with a standard basic dimension for the construction of houses. It is found in Japanese architecture where woven straw mats, called *tatami,* are a nominal 1-yard-by-2-yards and form the unit of measurement for a house. The modern architect Le Corbusier came up with his own standard unit, called a *modulor,* and used it as the scale by which all his designs were governed.

For today's homes the concept has less to do with units of measurement and more to do with basic building blocks of space—called modules—that either form a small house on their own or go on to form a larger unit *en masse.* This means that manufacturers are able to produce a variety of styles and floor plans for just about every budget—from modest starter home to mansion.

The modular home industry is structured much like the automobile industry, only on a smaller scale. Each company has a network of independent local builder/dealers whose responsibility is to act as the "front-line" company representative. Typically a builder/dealer will help you select a home style and floor plan that fits your needs and pocketbook and then turn in your order to the factory. There each house is produced exactly as ordered, complete with wall coverings, carpet, appliances, light and plumbing fixtures, hardwood floors, ceramic tile, kitchen cabinets, and other options. Because modular builders buy their materials in quantity, they are not likely to offer a wide range of interior or exterior finishes or appliances. But most of the companies are small enough that they are willing to order and install items that you want—provided you are willing to pay the extra price. This means that if you really feel you must have a carpet upgrade that is not standard, the manufacturer will put it in at an added cost.

Usually, while the factory is busy building your house, the local builder/dealer is busy installing your foundation. When the house arrives, either the builder/dealer or a factory team "sets" the house on its foundation, using a heavy crane to put the modules in place. Then the local tradespeople hook up your water; electricity; heating, venting, and air-conditioning (HVAC) system; and everything is ready for you to move in.

Cost: Because they're built on a production line, modular homes can save you money. Given the same lot and similar floor plan, style, finish, and builder profit, a modular house will typically cost you about $10,000 less than its site-built equal (perhaps even more depending on where you live). And if you're willing to take on some of the finish work yourself, you can save even more money. Some manufacturers will work with you to delete some of the operations they do in the factory (like painting, wallcovering, and trim). And some also have models that come roughed in so that you can do the finish work yourself (such as insulation, wallboard, plumbing and lighting fixtures, trim, and so on).

Because a modular home can be built and erected in less time, you can also realize savings in interest costs on a construction loan. For example, you would likely need only 30 days to finish a modular house (once it has been erected on your lot), while a site-built would take an average of 120 to 150 days to complete. Given a $60,000 construction loan at 12 percent interest, you therefore realize a net savings of $5,400 ($600 for 30 days versus $6,000 for 120 days).

Size and Style: Most modular manufacturers offer products in a wide range of styles and sizes. Today you can get a 1,000-square-foot starter home, a 2,000-square-foot colonial, a 2,500-square-foot ranch, or a 5,000-square-foot architect-designed, split-level contemporary—all modular. And according to one home builder who recently switched to modular building, the only difference between modular homes and the traditional style of construction is that a hundred trips are needed to deliver the material to the site of a custom-built home but only one trip is required to drop off the modular units.

Quality: Because they are built at a common location, modular homes enjoy a constant labor force working within a controlled environment. Materials such as kiln-dried lumber can be ordered in quantity and thus quality-controlled. Special tools and jigs assure accurate dimensioning. Horizontal tables allow gypsum wallboard to be laid flat on wall panels so that it can be glued and screwed (virtually impossible under field conditions). And because most modulars today are still built using the traditional 2-by-4 wood framing method they can be easily maintained or altered using locally available materials and labor.

Time: Most modulars leave the factory 95 percent complete. This means that by the time your house arrives at your lot, it already has plumbing and electrical fixtures installed and ready for hook-up. The walls are painted, wallpapered, or both. Ceramic tile and wood flooring is installed. All appliances are in place and ready to be turned on.

Assuming you already have a lot on which to build, a typical scenario might look like this:

Step	Activity	Time
1	Factory puts your house into the production schedule	30 days
2	Factory builds your house	7 days
3	Factory ships house to site	1 day
4	Factory erects house on site	1 day
5	Dealer finishes house on site	14 days
	Total	53 days

A.

B.

A. Nationwide Homes set a record by constructing this 2,300-square-foot modular home in less than twenty-four hours. Here, the first section is being lowered onto the foundation just fifteen minutes after it arrived on the site.

B The back portion of the lower level placed thirty-five minutes after start.

C. The upper level takes shape two hours after start.

C.

D.

E.

F.

G.

D. Two hours and twenty-five minutes later the roof is assembled.

E. A crane sets the prefinished gable.

F. The crew is under roof only three hours after start.

G. The finished product—19 hours and forty-two minutes later. If this were a site-built home, it would have taken between twenty-five and thirty weeks to construct.

As you can see, the biggest chunk of time is used by the factory to fit your house into its production schedule. Normally your builder/dealer would use this time to prepare the site (clearing, grading, building the foundation) so it would be ready to receive your house from the factory.

Financing: Modular homes are appraised and financed just as site-built houses are. Most lenders are pleased to include modular homes in their loan portfolios and seek them out as actively as they would any other type of home lending.

Normally you would go through the following procedure to obtain financing for a modular house:

1. Obtain a construction loan for building the house. This would include the COD factory charge, site preparation costs, and any additional site work you might require (concrete driveway, deck, porch, patio, landscaping). Typically, lenders ask that you put 10 percent down in applying for the loan, which you can use as a down payment to the factory as an order to build.
2. Use your construction loan to pay the COD charges when the house is delivered to the site and to pay your builder/dealer.
3. Obtain permanent financing for your completed house either from your construction lender or from another lender. (Note: Modular housing is looked upon favorably by the secondary mortgage market. Depending on which route you take, you should be able to obtain a conventional mortgage or one from the Department of Veterans Affairs (VA) or Federal Housing Administration (FHA) with little or no problem).

Codes: All manufacturers build to one of three model codes in general use in the United States:

- Uniform Building Code (UBC), used by over 1,200 jurisdictions, is almost the exclusive code west of the Mississippi.
- Building Officials and Code Administrators (BOCA) is used primarily in the Northeast.
- Standard Building Code, published by the Southern Building Code Congress International (SBCCI), is used almost exclusively in the Southeast.

In addition, manufacturers work closely to comply with state and local regulations in the areas to which they ship their products. This means all framing and rough inspections are approved *before* the house is shipped to the job site. The only inspections that need to be done once the house is set are the building and mechanical finals that assure that the home is safe for occupancy.

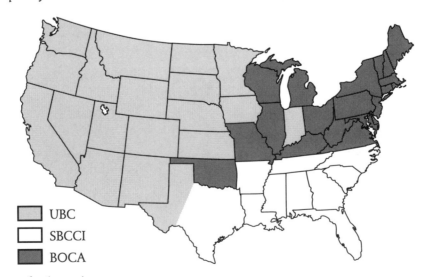

UBC
SBCCI
BOCA

Regional breakdown of code regulators.

Choosing a Modular Home Manufacturer: The first thing you will want to note about modular builder/dealers is what geographical area(s) they serve. Since modules are moved over the roadway by trucks, manufacturers are limited in their radius of distribution. Some have gotten around this by setting up regional plants. Others stay where they are and ship products within their area.

Most companies give guaranteed delivery dates. Be sure yours does. When your builder/dealer has your foundation ready, you both will want your home to be there.

Modular Builders Directory

Since most companies offer a full range of styles and sizes of homes, including ranches, split-levels, two-stories, capes, contemporaries, and vacation homes, or will custom-build to your specifications, the most important item to look for in the following list is which companies serve the area where you live. Then you might want to write or call all those who do serve your area and get their sales literature. At the same time you can find out if they have a builder/dealer network. As most do, this means there is probably either a model or spec home nearby that you can check out.

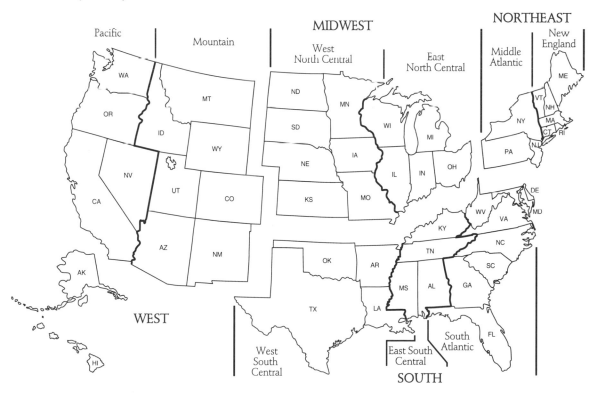

This map shows the geographic regions covered by modular builders. It is best to try to find a dealer in your region.

A.S.I. of New York, Inc., 5911 Loomis Road, Farmington, NY 14425. (716) 924–7151. Available in New England, New York.

Active Homes Corporation, 7938 South Van Dyke Road (M-53), PO Box 127, Marlette, MI 48453. (517) 635–3532. Fax (517) 635–3327. Available in Michigan, Indiana, Ohio.

All American Homes, Inc., 1418 South Thirteenth Street, Decatur, IN 46733. (219) 724–9171 or (219) 875–2421. Available in East North Central.

Alouette Homes, PO Box 187, Newport, VT 05855. (514) 539–3100. Fax (514) 539–0335. Available in Canada (Quebec, Ontario, New Brunswick, Nova Scotia) and New England.

Avis Homes Co., Henry Street, Avis, PA 17721. (800) 233–3052 or (717) 753–3700. Available in Northeast, Michigan, Ohio.

Benchmark Communities, Inc., 630 Hay Avenue, Brookville, OH 45309. (513) 833–4091. Available in Michigan, Indiana, Illinois, Ohio, Kentucky.

Canadian Portable Structures, 3901 Harrison Court, Burlington, Ontario, Canada L7R 3X4. (416) 335–5500. Available in Ontario, Quebec, New York.

Cardinal Homes, Inc., PO Box 10, Wylliesburg, VA 23976. (804) 735–8111. Available in Delaware, Maryland, West Virginia, Virginia, North Carolina.

Coker Builders, Inc., Box 8, Turbeville, SC 29162-0008. (803) 659–8585. Available in North Carolina, South Carolina, Georgia.

Contempri Homes, Inc., Stauffer Industrial Park, Taylor, PA 18504. (717) 562–0110. Available in Northeast.

Customized Structures, Inc., PO Box 884, Plains Road, Claremont, NH 03743. (603) 543–1236. Available in New England and New York.

Deluxe Homes of Pennsylvania, Inc., 499 West Third Street, PO Box 323, Berwick, PA 18603. (717) 752–5914. Available in Northeast.

Design Homes Inc., PO Box 411, West Fifth Street, Mifflinville, PA 18631. (717) 752–1001. Available in Northeast.

DKM Building Enterprises, Princeton Pike Corporation Center, Lawrenceville, NJ 08648. (609) 895–1433. Available in Northeast.

Dynamic Homes, Inc., PO Box 1137, Detroit Lakes, MN 56502. (218) 847–2611. Available in West North Central and Wisconsin.

Epoch Corporation, Route 106, PO Box 235, Pembroke, NH 03275. (603) 225–3907. Available in New England.

Excel Homes, Inc., RR #1, Box 27-C, Mount Pleasant Mills, PA 17853. (800) 345–6767 or (717) 539–2356. Available in Northeast, West Virginia, Maryland, Delaware, Virginia.

The Future Home Technology, PO Box 4255, 33 Ralph Street, Port Jervis, NY 12771. (914) 856–9033. Available in Northeast, West Virginia, Virginia, Maryland.

General Housing Corporation, 900 Andre Street, Bay City, MI 48706. (517) 684–8078. Available in Michigan, Ohio, Indiana, and Ontario, Canada.

Golden West Homes, 1308 Wakeham, Santa Ana, CA 92705. (714) 835–4200. Available in California, Arizona, Nevada, Hawaii.

Haven Homes, Inc., PO Box 178, Route 150, Beech Creek, PA 16822. (717) 962–2111. Available in Northeast, West Virginia, Delaware, Maryland, Virginia.

Harrington Homes Inc., Route 14, PO Box 98, East Montpelier, VT 05651. (802) 479–3625. Available in New England.

Heckaman Homes, PO Box 229, 26331 US 6 East, Nappanee, IN 46550. (219) 773–4167. Available in Michigan, Ohio, Illinois, Indiana.

H.H.I., 6520 East Eighty-second Street, Indianapolis, IN 46250. (317) 842–1900. Available in Michigan, Indiana, Ohio, Illinois.

Huntington Homes, Inc., PO Box 98, Route 14, East Montpelier, VT 05651. (802) 479–3625. Available in New England and New York (Eastern).

Integri Homes, 2320 Montclair, Leesburg, FL 32748. (904) 787–2056. Available in Florida.

Kaplan Building Systems, Inc., Route 443 East, PO Box 247, Pine Grove, PA 17963. (717) 345–4635. Fax (717) 345–2642. Available in Northeast.

Kings Haven, Inc., PO Box 228, Fort Payne, AL 35967. (205) 845–3550. Available in South.

Mitchell Brothers, Inc., 960 Forestdale Boulevard, Birmingham, AL 35214. (205) 798–2020. Available in Alabama, Mississippi, Georgia, Florida, Tennessee.

Mod-U-Kraf Homes Inc., PO Box 573, Rocky Mount, VA 24151. (703) 483–0291. Available in Southern Middle Atlantic, Eastern Kentucky and Tennessee and South Atlantic (except South Carolina, Georgia, Florida).

NRB, Inc., 115 South Service Road, PO Box 129, Grimsby, Ontario, Canada L3M 4G3. (416) 945–9622. Available in Ontario, Quebec, New York, and Pennsylvania.

Nanticoke Homes, Inc., PO Box F, Greenwood, DE 19950-0506. (302) 349–4561 or (302) 856–3340. Available in New Jersey, Pennsylvania, Maryland, Delaware, Virginia.

Nationwide Homes, 1100 Rives Road, PO Box 5511, Martinsville, VA 24115. (703) 632–7101. Available in New Jersey, Pennsylvania, Kentucky, Tennessee, and South Atlantic except Florida.

New Century Homes/Signature Building Systems, PO Box 9, 1100 West Lake Street, Topeka,

IN 46571. (800) 777–6637 or (219) 593–2962. Available in Michigan, Illinois, Indiana.

New England Homes, Inc., 270 Ocean Road, Greenland, NH 03840. (603) 436–8830. Available in New England.

North American Housing Corporation, 4011 Rock Hall Road, PO Box 145, Point of Rocks, MD 21777. (301) 694–9100. Available in Northeast, Kentucky, Tennessee, and South Atlantic except Florida.

Penn Lyon Homes, Inc., Airport Road, PO Box 27, Selinsgrove, PA 17870. (717) 374–4004. Fax (717) 374–6053. Available in Northeast and South.

Princeton Homes Corporation, PO Box 2086, Danville, VA 24541. (804) 797–3144. Available in Virginia, West Virginia, North Carolina, South Carolina.

Randal Company, Box 337, Piketon, OH 45661. (614) 289–4770. Available in Ohio, Kentucky, West Virginia.

Ritz-Craft Corporation, PO Box 70, Mifflinburg, PA 17844–0070. (717) 966–1053. Available in Northeast.

Schult Homes Corporation, 30 Industrial Park Road, Milton, PA 17847. (717) 742–8521. Available in Northeast.

Stratford Homes Limited Partnership, PO Box 37, 402 South Weber, Stratford, WI 54484. (800) 448–1524. Fax (715) 687–3453. Available in East North Central, Minnesota, Iowa.

Structural Modulars, PO Box 607, Southern Avenue, Clarion, PA 16214. (814) 764–5555. Available in Middle Atlantic, Virginia, West Virginia, Maryland, Delaware, Ohio.

Taylor Homes, Highway 71 North, PO Box 438, Anderson, MO 64831. (417) 845–3311. Available in Missouri, Illinois, Kansas, Arkansas, Iowa, Oklahoma.

Terra Quest Homes, Inc., 36696 Sugar Ridge Road, North Ridgeville, OH 44039–3832. (216) 327–6969 or (216) 362–7770. Available in Ohio, Michigan.

Terrace Homes, 301 South Main, Box 1040, Adams, WI 53910. (608) 339–7888. Available in Midwest.

Unibilt Industries, Inc., PO Box 373, Vandalia, OH 45377. (800) 777–9942 or (513) 890–7570. Available in Michigan, Indiana, Ohio, Kentucky, West Virginia.

Unicoast Industries, Inc., 2570 NW 16th Boulevard, Okeechobee, FL 34972. (813) 763–7787. Available in Florida.

Virginia Homes Manufacturing Corp., PO Box 410, Boydton, VA 23917. (804) 738–6107. Fax (804) 736–6926. Available in Northeast and South Atlantic.

Wausau Homes, PO Box 8005, Wausau, WI 54402–8005. (715) 359–7272. Available in East North Central, Minnesota, Iowa.

Westchester Modular Homes, PO Box 900, Dover Plains, NY 12522. (914) 832–9400. Available in Northeast.

Willcraft Home Systems—Division of Joe Williams Homes, Inc., PO Box 2135, Newnan, GA 30264. (404) 251–4655. Available in Georgia, Alabama.

4

Mobile Homes

A four-bedroom mobile home replete with two-car garage.

If you're looking for a new home and have Cadillac tastes but a Chevy budget, a mobile home may be right for you.

Mobile homes—the industry likes to call them *manufactured housing*—have always been the poor stepchild in the home-building business. But the kid's coming of age.

Historically mobile homes met the immediate needs of thousands of returning World War II veterans who found they were unable to afford anything but the most basic kind of housing. But the industry has since grown. Today mobile home production is booming, up 60 percent since 1980, and white-collar workers are the fastest-growing segment of new buyers. In fact, *The Charlotte Observer* says about North Carolina, one of the states where mobile home living is popular: "If trends of the 1980s continue, manufactured homes will outnumber traditional site-built single-family homes in North Carolina twenty-one years from now."

The term *mobile home* is actually a misnomer. Manufacturers are required by law to include a metal chassis as an integral part of the floor system to which removable axles and wheels are attached for transporting the unit. But when today's mobile home reaches the job site, it's most often placed on a permanent foundation, thereby becoming almost indistinguishable from a site-built house.

The real glitch in the popularity of mobile homes comes from their public image. Even today most U.S. communities have zoning that restricts mobile home locations, thus preventing such units from being used in place of more conventional housing.

Two states that don't have such restrictions are Florida and California. In California the big news is the shift from senior citizen-oriented mobile home projects to the family market. And as the market changes, both mobile homes and lots are getting bigger, with three-bedroom homes now the norm, along with site-built, two-car garages. Homes on these lots

		Ave. Sales Price	Cost/sq. ft.
Single-section 50' to 80' / Floor Plan / 14' to 16' / 970 sq. ft.		$19,200	$19.79
Multi-section 50' to 60' / Floor Plan / 24' to 36' / 1,440 sq. ft.		$34,800	$24.17
Site-Built 50' / Floor Plan / 40' / 2,000 sq. ft.		$106,500 (not incl. land)	$53.25

are being designed with many of the same features as homes in conventional subdivisions. For example, some models face the street, and others use such traditional building materials as stucco, lapped siding, and roofing shingles. Additionally, at least one new California project is slated to feature two-story units, the first of their kind.

Cost: If price is a big consideration, mobile homes have a big advantage. Take a look at the table.

Size and Style: The manufactured housing industry produces homes in complete modules referred to as either *single-wides* or *multi-sectionals.* Single-wides are complete, self-contained units and are delivered to the job site ready to be installed and hooked-up. Multi-sectionals are delivered in sections and, like modular homes, must be fastened together to form one complete unit.

Today's mobile home definitely does not fit the old stereotype. Styles range from traditional to contemporary and floor plans can be had in a variety of configurations, with some exceeding 2,000 square feet. Some manufacturers have recently introduced such features as lap siding, shingled roofs, custom-style windows, kitchens with custom appliances, bathrooms with sunken tubs, cathedral ceilings, and even wood-burning fireplaces.

Mobile homes can have spacious interiors just like those in site-built homes.

Quality: Since June 15, 1976, all manufactured housing has been built to a national building code established by the U.S. Congress and administered by the U.S. Department of Housing and Urban Development (HUD). This code regulates the design, construction, strength, durability, fire resistance, energy efficiency, and installation and performance of the heating, plumbing, air-conditioning, thermal, and electrical systems of all manufactured homes.

Consequently, when you walk onto a mobile home dealer's lot, you can be guaranteed that each home was built in accordance with a federal code and was inspected by federal inspectors (or their approved substitutes). That doesn't mean that all products are the same. But it should assure you that the manufacturer didn't cut any corners affecting the home's code-worthiness.

Mobile home builders are known, however, to be "nickel-squeezers." It's what gives them their price superiority over other forms of housing. They'll scrimp on some areas of framing or sheathing that a modular builder won't. But this doesn't necessarily mean a flimsier product—only a cheaper one. And there are some features that you will get in a mobile home that you won't get in one built to site codes. For example, you'll always have an exit via an escape window. Most city codes don't require one. And you'll also find noncombustible material around furnaces and ranges. Not all city codes specify these niceties.

Time: Like modular homes, mobile homes leave the factory almost complete. Walls and ceilings have been painted, carpeting has been installed, and appliances and fixtures are in place and ready to be turned on once the major hook-ups have been made.

Financing: Most buyers choose to finance directly through a dealer under a retail installment contract. Bank and savings-and-loan institutions will also provide financing, but at terms likely to be less favorable than those for site-built homes. If you buy a mobile home, for example, you will likely end up paying around three percentage points on top of the going rate for a thirty-year mortgage (though the term of your mortgage isn't likely to exceed fifteen years).

Lenders say the differences in rates are due to the fact that smaller loans cost more to administer. In addition, they say they have to spend more time on collection efforts to keep such loans current. Also, they must pay higher costs themselves to primary lenders for their manufactured home funds.

When a mobile home is permanently attached to a foundation and sold with land, it generally may be financed with a real estate mortgage for up to thirty years at a fixed rate or with any of the variable or adjustable rate mortgages that are available to finance site-built homes. Programs available through the VA and FHA to guarantee loans or insure loans on manufactured homes are available to qualified buyers. Certain foundation requirements must be met, however: Wheels, axles, and hitches must be removed, and the home and land must be treated as a single real estate entity under state law.

A.

B.

C.

D.

E.

F.

G.

A. Here's a look inside a mobile home factory. First, the metal chassis is welded to form the platform for a new mobile home.

B. Next, the floor system is installed over the chassis.

C. The carpeting is laid in next.

D. Exterior walls are added.

E. Interior walls and wall covering soon follow.

F. The home starts to take shape as the roof is installed.

G. The completed home is ready for shipping.

The Government National Mortgage Association (GNMA) has a secondary market program for traditional manufactured home personal property loans. But only loans guaranteed by the VA or the FHA Title I program are eligible for the GNMA program.

The Federal National Mortgage Association (FNMA) and the Federal Home Loan Mortgage Corporation (FHLMC) also have secondary market programs for mobile homes treated as real estate—that is, such homes must be permanently affixed to a suitable foundation, and the home and land must be considered real estate under state law.

Codes: All mobile home manufacturers build their homes in accordance with the HUD Code discussed on page 000.

Choosing a Mobile Home Manufacturer: There are nearly one hundred companies building manufactured homes in about 250 factories throughout the United States. In 1989, manufactured home sales amounted to almost $5.3 billion.

In this book we've listed the thirty-nine major manufacturers that either have their own dealer network or sell their homes through independent dealers or through individuals who operate trailer parks (now called *land-lease communities*). Additionally, there are some residential developers who are looking to the mobile home as an affordable alternative to the traditional site-built house.

Normally dealers assume the role of a general contractor and put house and location together. They make sure the house meets the specifications of local law, that utilities and city or well water are ready for hook-up, and that the foundation is in place and ready for installation.

Select a dealer as carefully as you would the home itself. Most are reputable but there have been exceptions. Ask questions, get a list of references, and ask around at local government and business reference agencies like the chamber of commerce or Better Business Bureau.

If you intend to place your home on private property, you should check with your local municipality (city, county, village, or township) and find out what regulatory issues may apply to mobile homes (zoning, appearance standards, installation requirements). As of July, 1986, fourteen states—California, Colorado, Florida, Indiana, Iowa, Kansas, Maine, Minnesota, Nebraska, New Hampshire, New Jersey, Oregon, Tennessee, and Vermont—had passed legislation that prohibits exclusion and unfair regulatory treatment of mobile homes. If you run into problems, you can probably get help from the manufacturer, retail dealer, or state manufactured-housing association (listed at the back of this book).

If you plan to go into a land-lease community, you will likely want to consider such matters as location, rent, rules, special facilities, and restrictions. For locations you should probably contact the state association or consult your local yellow pages under "Mobile Homes—Parks." Other sources of information include retail dealers and state regulatory agency.

Mobile Home Builders Directory

The following companies are Manufacturer Members of the Manufactured Housing Institute and represent about 65 percent of the total mobile home production in the United States. As mentioned earlier, you can't buy directly from any of them and will likely have to go through a retail dealer or other intermediary. If you need help in finding an outlet, you can write or call the manufacturer and a company representative will put you in touch with the right party.

American Family Homes, Inc., Highway 71 North, PO Box 438, Anderson, MO 64831. (417) 845–3311. Fax (417) 845–3315.

Bayshore Homes of California, Inc., 11 North County Road 101, Box 1427, Woodland, CA 95695. (916) 662–9621. Fax (916) 661–1179.

Bellcrest Homes, Inc., 206 Magnolia Street, PO Box 603, Millen, GA 30442. (912) 982–4000. Fax (912) 982–2992.

Brigadier Homes of North Carolina, Inc., PO Box 1007, Highway 64 East, Nashville, NC 27856. (919) 459–7026. Fax (919) 459–7529.

Buccaneer Homes of Alabama Inc., PO Box 1418, Hamilton, AL 35570. (205) 921–3135. Fax (205) 921–7390.

Burlington Homes of New England, PO Box 268, Route 26, Oxford, ME 04270. (207) 539–4406. Fax (207) 539–2900.

Cavalier Homes, Inc., PO Box 300, Highway 41 North, Addison, AL 35540. (205) 747–1575. Fax (205) 747–2107.

Champion Home Builders Company, 5573 North Street, Dryden, MI 48428. (313) 694–3195. Fax (313) 796–2145.

Chief Industries, Inc. Bonavilla Homes, PO Box 127, West Highway 34, Aurora, NE 68818. (402) 694–5250. Fax (402) 694–5873.

Clayton Homes, Inc., PO Box 15169, Knoxville, TN 37901. (615) 970–7200. Fax (615) 970–1238.

The Commodore Corporation, PO Box 577, Goshen, IN 46526. (219) 533–7100. Fax (219) 534–2716.

Crestline Homes, Route 3, Box 67, Laurinburg, NC 28352. (919) 276–0195. Fax (919) 276–7989.

Fisher Corporation, PO Box 1000, Highway 52, Richfield, NC 28137. (704) 463–1341. Fax (704) 463–5199.

Fleetwood Enterprises, 3125 Myers Street, PO Box 7638, Riverside, CA 92523. (714) 351–3500. Fax (714) 351–3776.

Franklin Homes, Inc., Route 3, Box 207, Russellville, AL 35653. (205) 332–4510. Fax (205) 332–5449.

Fuqua Homes, Inc., 7100 South Cooper, Arlington, TX 76017. (817) 465–3211. Fax (817) 465–5125.

Gateway Homes, PO Box 728, Guin, AL 35563. (205) 468–3191. Fax (205) 468–3336.

Golden West Homes, Inc., 1308 East Wakeham, Santa Ana, CA 92705. (714) 835–4200. Fax (714) 835–6232.

Guerdon Industries, Inc., 5285 South West Meadows, Suite 315, Lake Oswego, OR 97035. (503) 624–6400. Fax (503) 620–5929.

HomeCorp, Inc., 19224 C.R. #8, Bristol, IN 46507. (219) 848–4421. Fax (219) 848–5755.

Keiser Homes of Maine, PO Box 470, Oxford, ME 04270. (207) 539–8883. Fax (207) 539–4446.

Kit Manufacturing Company, PO Box 848, Long Beach, CA 90801. (213) 595–7451. Fax (213) 426–8463.

Mansion Homes, PO Box 39, Plank Road, Robbins, NC 27325. (919) 948–2141. Fax (919) 948–3752.

Mobile Home Estates, Inc., Route 4, Bryan, OH 43506. (419) 636–4511. Fax (419) 636–9144.

Moduline International, Inc., PO Box 3000, 205 College Street, SE, Lacey, WA 98503. (206) 491–1130. Fax (206) 491–1135.

Oakwood Homes Corporation, PO Box 7386, Greensboro, NC 27417–0386. (919) 855–2400. Fax (919) 852–1537.

Patriot Homes, Inc., 57420 C.R. #3, South, Elkhart, IN 46517. (219) 293–6507. Fax (291) 522–2339.

Peach State Homes, PO Box 615, Adel, GA 31620. (912) 896–7420. Fax (912) 896–2575.

R-Anell Custom Homes, Inc., PO Box 428, Denver, NC 28037. (704) 483–5511. Fax (704) 483–5674.

Redman Homes, Inc., 2550 Walnut Hill Lane, Suite 200, Dallas, TX 75229. (214) 353–3600. Fax (214) 956–9986.

Rochester Homes, Inc., East Lucas Street, PO Box 587, Rochester, IN 46975. (219) 223–4321. Fax (219) 862–2239.

Sun Belt Energy Housing, Highway 5 North, PO Box 340, Haleyville, AL 35565. (205) 486–9535. Fax (205) 486–4197.

Schult Homes Corporation, PO Box 151, 221 US 20 West, Middlebury, IN 46540. (219) 825–5881. Fax (800) 955–2355.

Skyline Corporation, PO Box 743, Elkhart, IN 46515. (219) 294–6521. Fax (219) 293–7574.

Sunshine Homes, Inc., PO Box 507, Red Bay, AL 35582. (205) 356–4427. Fax (205) 356–9694.

Virginia Homes Manufacturing Corp., PO Box 410, Boydton, VA 23917. (804) 738–6107. Fax (804) 738–6926.

Victorian Homes, Inc., PO Box 707, 11948 C.R. #14, Middlebury, IN 46540. (219) 825–5841. Fax (219) 825–9851.

Westway Homes, Inc., PO Box 3850, Ontario, CA 91761. (714) 947–3816. Fax (714) 947–2307.

Wick Building Systems, Inc., 404 Walter Road, PO Box 490, Mazomanie, WI 05360–0490. (608) 795–4281. Fax (608) 795–2740.

5

Panelized Homes

A panelized wall section is moved into position.

Panelized homes occupy the middle ground between modular and site-built. When a modular home leaves the factory, it is 95 percent complete. Depending on the type of construction, a panelized house is 30 to 60 percent complete. When it arrives at the job site, it can be erected in a fairly short period of time and is then finished in the same manner as a site-built home.

The advantage to using panels is the flexibility buyers have in choosing a style and floor plan. With factory-crafted wall panels, home shapes can be—and often are—unique and attractive. Complex roof systems, interior cathedral ceilings, lofts, and multiple floor levels are quite common. Because the panelized concept permits virtually *any* house design to be constructed, many panel manufacturers not only offer a broad range of standard plans but also are able to take other house plans and produce the prefabricated components from which the home can then be assembled. Many have an architectural team on board that can help buyers come up with a design, should they not have something already in mind.

For a buyer, a panelized home has the normal factory-crafted benefits of tight tolerances, quality materials, high standard of workmanship, and reduced construction time. For a builder, panelized homes are predictable and simplified systems that go up quicker and cause less waste. And because the whole building shell is supplied, a builder doesn't have to keep going to the lumberyard for materials.

Among the earliest examples of this factory-crafted concept were the houses that came from Sears, Roebuck and Company. In the early 1900s nearly 100,000 families turned to Sears not only for everything to fill an American home, but for the home itself. Sears' customers ordered homes by mail and received them by rail. The top-of-the-line model featured high-quality exterior wood sheathing and siding, cedar shingles, and yellow pine, oak, or maple flooring and inside trim. Even modest, one-story homes were available for use as summer cottages.

Many Sears homes still survive in small towns and big cities throughout the United States. For information about their locations, contact the National Trust's *Historic Preservation* magazine at (202) 673–4000.

Today's panelized homes, which are high-tech updates of the Sears concept, belong to three categories: (1) open wall, (2) closed wall, and (3) precut.

Open Wall: Open-wall manufacturers supply you with a kit or package of two-dimensional components, including wall, floor, and roof panels, partitions, and plumbing walls. The components are made up in the factory and connected to each other on the job site. The sizes of the panels will vary according to the manufacturer and may be anywhere from 2 feet to 42 feet in length. When they arrive at the site they are assembled to form the house's shell. This shell is then finished in the same manner as the more traditional site-built house.

Closed Wall: Closed-wall manufacturers ship wall sections that need almost no on-site finishing. Typically, these panels include exterior sheathing, insulation, wiring, and gypsum wallboard. Plumbing is handled by providing factory-completed modular bathrooms, kitchen, utility/laundry, and furnace/equipment rooms that can be "plugged" into the house by tying in their lines with those already run in the house. This method speeds up the finishing process, and most houses featuring a closed-wall system finish out in far less time than those using the open-wall method.

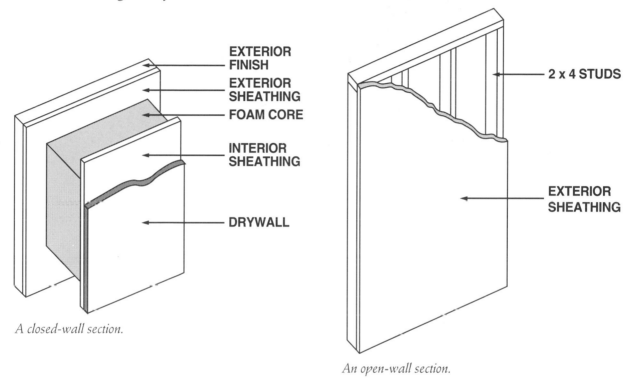

EXTERIOR FINISH
EXTERIOR SHEATHING
FOAM CORE
INTERIOR SHEATHING
DRYWALL

2 x 4 STUDS

EXTERIOR SHEATHING

A closed-wall section.

An open-wall section.

Precut: Precut homes are a borderline method of panelized building. As Sears did with its early models, precut manufacturers normally furnish studs, joists, and other structural members to size, predrilling and notching them. They may also finish wall panels, tongue-in-groove flooring, and other components. They then package them in assembly sequence and deliver them to the job site. The precut method permits a wider variety of house plans and styles, including post-and-beam and barn structures. Like open-wall houses, precut structures require the same finishing sequence as a site-built house.

Cost: Because they rely on a system in which most or all of the components are factory-supplied, panelized homes can save you money on building time and materials. But it's unfair to say that as a class they cost less than a house of similar style and square footage that's site-built. Your best bet is to compare "apples with apples." For example, if a manufacturer can provide you with an architect-designed, panelized home with top-of-the-line compo-

These two homes show the stylistic range of the panelized system.

nents and high-quality materials, then you'd have to ask yourself how much a similar home would cost if you did it on your own—that is, hired an architect, bought all the materials and components, hired a precision carpenter, and so on. Normally you're still ahead of the game, financially, if you go factory-crafted. And you still save a lot of time in both the design and the building process. Just be alert to the degree of finish each manufacturer's product requires before calculating your total savings.

Size and Style: This is where panelized homes run circles around either modular or mobile homes. Because they rely on smaller components, panel builders can offer you a wider variety of configurations. (The difference is like that of building a house out of Legos® versus building one out of boxes.) Today you can choose colonial, salt box, polygonal, Victorian, barn, contemporary, farm, medieval—in fact, almost any style and floor plan you can imagine. And if you don't see what you like in a company's plan book, chances are the people there can work from your plans or help you draw up plans that you *do* like.

Quality: Like other forms of factory-crafted products, panelized homes are built at a common location and employ a constant labor force working within a controlled environment. And because they deal in large quantities, panel builders can order a higher grade of wood than you're likely to find at the local lumberyard. This means nice, straight, almost knot-free framing material that is probably kiln-dried and therefore less likely to warp or shrink over time (meaning better-looking interior walls and ceilings).

Some panel builders also offer such niceties as laminated fir posts and beams for interiors and clear cedar trim and siding for exteriors. These materials are almost impossible to get at the local level and add greatly to a home's aesthetic and actual value. Other features offered by many panel manufacturers include solid wood interior doors, polished brass hardware, and forged steel connector plates to join structural elements.

Time: Open-wall and precut homes take longer than a closed-wall panel house to erect and finish. In addition, an even larger amount of time is needed for certain complex house designs. Your best bet is to go with the manufacturer's estimate on package erection and then factor in such local conditions as weather patterns and labor supply. As you would with other forms of factory-crafted homes, count on a good chunk of time for the factory to fit your house into its production schedule. This is especially true if yours is a custom design.

Financing: Panelized homes are appraised and financed just like those that are site-built. You may find that some local lenders are not aware of the concept and are unwilling to give you money. But in most cases you should have no problem if you show them your plans plus photos of models or completed homes (which the manufacturer or local builder/dealer can supply). Following the same procedure should enable you to solve any problems you may run into with a local development with deed restrictions or architectural review.

Codes: Since panelized builders ship their products around the nation (and in some cases around the world), they have designed their products to meet or exceed any national, state, or local building codes. If there's any question about a particular product's ability to meet such codes, check with the manufacturer or its local builder/dealer representative before you buy. Either one should know.

This sequence shows the stages of panelized building. After the foundation is poured, the home is erected using factory-crafted panels.

Choosing a Panelized Builder: Like other factory-crafters, panelized builders have a network of builder/dealer representatives who operate as middlemen between you and the factory. Through them you can see a model, discuss your needs in more detail, and order a new home.

Panelized Home Manufacturers Directory

Unlike modular builders, which are limited in their product distribution, panelized manufacturers can and do distribute their products worldwide. Because most panelized builders fall into one of three categories, we have identified each listing as to its main system: open wall, closed wall, or precut.

As with modular builders, you will probably want to write or call those who interest

Panelized homes can offer a wide range of interiors with open floor plans and contemporary window treatments.

you and get their sales literature. At the same time you can find out if they have a builder/dealer network and either a model or spec home nearby that you can check out.

Acorn Structures, Inc., PO Box 1445, Concord, MA 01742. (508) 369–4111. Fax (508) 371–1949. Open wall.

ALH Building Systems, US 224 West, PO Box 288, Markle, IN 46770. (219) 758–2141. Open and closed wall.

Amos Winter Homes, Inc., Glen Orne Drive, Brattleboro, VT 05301. (802) 254–6529. Closed wall.

Armstrong Lumber Co., Inc., 2709 Auburn Way North, Auburn, WA 99002. (206) 852–5555. Open wall.

Bristye, Inc., Box 818, Mexico, MO 65265. (314) 581–6663. Open wall.

Carolina Builders Corp., PO Box 58515, Raleigh, NC 27658. (919) 750–8280. Open wall.

Classic Post & Beam Homes, PO Box 546, York, ME 03909. (800) 872–BEAM. Precut, closed wall.

Coastal Structures, Inc., PO Box 631, Laurence Road, Gorham Industrial Park, Gorham, ME 04038. (207) 854–3500, (800) 341–0300 (New England only), (800) 442–6363 (Maine only). Open wall.

Crest Manor Homes, Inc., PO Box 884, Martinsburg, WV 25401. (304) 267–4444. Open wall.

Deltec Homes, 604 College Street, Asheville, NC 28801. (800) 642–2508. Fax (704) 254–1880. Open wall.

ENDURE-A-LIFE-TIME Products, Inc., 7500 North West Seventy-second Avenue, Miami, FL 33166. (305) 885–9901. Closed wall.

Enstar Building Systems, Inc., 1207 East South Street, Orland, CA 95963. (916) 865–4117. Open wall.

Fischer Corporation, 1843 Northwestern Parkway, Louisville, KY 40203. (502) 778–5577. Closed wall.

Forest Home Systems, Inc., Route 522, RD #1, Box 131K, Selinsgrove, PA 17870. (800) 872–1492 or (717) 374–0131. Open wall.

Harvest Homes, Inc., 1 Cole Road, Delanson, NY 12053–0189. (518) 895–2341. Open wall.

K-K Home Mart, Inc., 420 Curran Highway, North Adams, MA 01247. (413) 663–3765. Open wall.

Korwall Industries, 326 North Bowen Road, Arlington, TX 76012. (817) 277–6741. Closed wall.

Lindal Cedar Homes, 4300 South 104th Place, PO Box 24426, Seattle, WA 98124. (206) 725–0900. Closed wall.

Miron Truss and Component Corporation, 54 Railroad Avenue, Albany, NY 12205. (518) 438–6811. Open wall.

Northern Counties Homes, Route 50 West, PO Box 97, Upperville, VA 22176. (703) 592–3232. Open wall.

Northern Homes, Inc., 51 Glenwood Avenue, Queensbury, NY 12804. (518) 798–6007. Open wall.

Pacific Southern Buildings, Inc., PO Drawer C, Covenington Road, Marks, MS 38646. (601) 326–8104. Open wall.

Pond Hill Homes, Ltd., Westinghouse Road, R.D. 4, Box 330-1, Blairsville, PA 15717. (412) 459–5404. Closed wall.

Riverbend Timber Framing, Inc., PO Box 26, 9012 East US 223, Blissfield, MI 49228. (517) 486–4355. Precut, closed wall.

Saco Homes, 21 West Timonium Road, Timonium, MD 21093. (301) 252–3030, (800) 695–SACO. Fax (301) 252–6164. Open wall.

Shelter Systems Corporation, PO Box 830, Westminster, MD 21157. (301) 876–3900. Open wall.

Timberpeg, PO Box 474, West Lebanon, NH 03784. (603) 298–8820. Precut, closed wall. Other plant locations in North Carolina, Colorado, Nevada.

Today's Building Systems, 195 Union Street, Suite E, Newark, OH 43055. (614) 345–3551. Open wall.

Unified Corporation, PO Box 87, Goochland, VA 23063. (804) 556–6275. Open wall.

Winchester Homes, Inc., 1321 Western Avenue, Baltimore, MD 21230. (301) 244–8112. Open wall.

Woodmaster Foundations, Inc., 845 Dexter Street, PO Box 66, Prescott, WI 54021. (715) 262–3655. Open wall.

Yankee Barn Homes, Inc., HCR 63, Box 2, Grantham, NH 03753. (603) 863–4545. Precut, closed wall.

Wausau Homes, Inc., PO Box 8005, Wausau, WI 54402–8005. (715) 359–7272. Open wall and closed wall.

Wick Homes, PO Box 818, Mexico, MO 65265. (314) 481–6663, (800) 892–5770. Open wall.

6

Log Homes—
Romancing the Home

A factory-crafted log home in the works.

If your idea of a log home is a small vacation cabin in the mountains, think again. According to log home producers, over 90 percent of all log homes sold in the United States today are for primary residences. And except for the rustic look that distinguishes log homes, these houses can be as spacious and contemporary as those built by other construction methods.

Owning a log home does carry with it a certain mystique. For some it's the look of natural wood inside and out. For others it's a conscious connection with America's pioneer past by using a building tradition that has remained largely unchanged for hundreds of years.

Log home producers are divided into two distinct groups: *handcrafters* and *manufacturers*.

Handcrafters produce logs for their homes in much the same way as the early settlers did. They go into the woods and select each tree with which they plan to build the house. They use hand tools to cut and shape each log, and as a result, their homes have a handcrafted appearance.

In contrast, log home *manufacturers* produce components in a factory. These factory-crafted houses generally cost less than their handcrafted counterparts. They are usually of a higher quality as well, because they are made under controlled conditions using expensive machinery that works to tight tolerances. They can be erected in less time than a handcrafted log home because they require less labor. And they greatly reduce waste on the job site.

Additionally, you can save a lot of money if you build a log home yourself using factory-crafted components. A survey conducted by the National Association of Home Builders noted that 75 percent of all log home buyers cut costs by helping build their own homes. Those who did such work paid on the average about $48 a square foot for their finished house, while those who didn't shelled out nearly $68 per square foot for their completed home.

Getting involved in the construction process doesn't have to mean hefting heavy logs or nailing shingles. One family, for example, saved $25,000 to $35,000 on their 1,800-square-foot log home by doing much of the site clearing, site preparation, and interior carpentry themselves. And this was their *first house.*

The appeal of log homes has resulted in a sizable industry devoted to their manufacture. Today more than 500 companies produce a total of 25,000 log homes annually, which represents a $3 billion investment. And the numbers are growing.

For many the beauty of log homes is in the elegant simplicity of the basic building system as the logs themselves take the place of a complex wall system of 2-by-4 studs, insulation, and interior and exterior sheathing used in most home building. The logs, like bricks, are placed in rows or "courses" to form structural walls that support themselves and a roof system. And since wood is such an excellent natural insulator, today's log homes also tend to be energy savers; tongue-and-groove fit, foam gaskets, and modern caulkings make them weather tight.

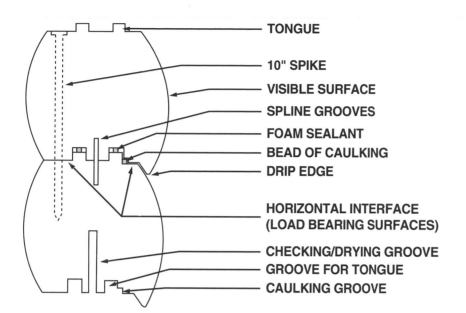

TONGUE

10" SPIKE

VISIBLE SURFACE

SPLINE GROOVES

FOAM SEALANT

BEAD OF CAULKING

DRIP EDGE

HORIZONTAL INTERFACE
(LOAD BEARING SURFACES)

CHECKING/DRYING GROOVE

GROOVE FOR TONGUE

CAULKING GROOVE

This illustration shows how logs are joined together.

Cost: Because they use a large number of (relatively) small components and require a good amount of site labor to complete, log homes generally compare in price with panelized structures. Normally, they are not sold as starter homes. Instead, they tend to be either move-up homes or large, custom, or "dream" homes. A rough rule-of-thumb to use in figuring the completed cost of a log home (package price plus erection and finish) is to multiply the square footage of a particular plan by $60.

Not only do building systems vary greatly from manufacturer to manufacturer, but package contents vary as well. There are basically three types of materials packages:

Walls only	$
Structural Shell	$$
Complete	$$$

Walls-only is the basic package. It contains logs for the exterior walls of the home, as well as fastening and weather-tight materials. The package may or may not include windows and doors, and the logs may or may not be precut. Although initially walls-only is the cheapest package, when you add in the roof structure and other components the total may not be the lowest.

A structural-shell package should contain everything found in a walls-only package, plus materials for a roof system, exterior doors and windows (including hardware), and anything else required to build a weather-tight, exterior shell. Again, logs may or may not be precut.

A complete package should have everything a structural package has, plus most of the other materials needed to erect a log home. These would include floor system, interior stud partitions, stairs, doors, hardware, and interior wall paneling. Logs may or may not be precut.

Size and Style: Deciding to build a log home doesn't mean you'll end up with something that looks like the log cabin Abe Lincoln grew up in. Most log home manufacturers produce a variety of standard plans. You can select or modify the one that suits your needs and your pocketbook. Many companies also have experienced designers on their team who can produce a custom log home from almost any floor plan you can imagine. Log manufacturers have built custom houses, condominiums, and churches. One manufacturer in North Carolina recently completed an entire retirement community.

Assembling factory-crafted components for a log home is similar to that of a stick-built home except that the logs fit together in a prearranged pattern to form the walls.

Quality: Log manufacturers will tell you they'll guarantee two things about a log home: (1) that logs will settle and (2) that logs will shrink. But not to worry. All wood (kiln dried and air dried) has some moisture, so most manufacturers have made provisions in their engineering and design for log settlement and shrinkage. But it would be worth your while when contacting individual manufacturers to ask what kind of logs they use (for instance, kiln-dried logs have less moisture content and therefore less settlement and shrinkage) and how their joinery system plans for the inevitable movement. Most log joinery systems feature high-tech caulks, chinking, and foam gaskets that ensure efficient weather tightness and minimum log movement.

Both the American Society for Testing and Materials (ASTM) and Timber Products Inspection Incorporated (TP) have developed grading programs that set minimum standards for log home manufacturers. The criteria by which they judge logs include slope of grain; knots; shakes, checks, and splits; straightness; scars; holes; and organic degradation (wood rot). Both programs are recognized by the three model-building code agencies. The North American Log Homes Council uses the ASTM criteria for its voluntary log grading standards program in which fifteen companies take part. Another forty-eight firms subscribe to the TP program.

Many reputable producers do not take part in either grading program, although their logs would meet or exceed the standards. The only way, however, that you can be assured that a producer's logs do meet the established standards is to find out if that producer takes part in a recognized log-grading program.

Time: Log homes tend to be as labor intensive as a site-built house and take about the same time to erect and finish. Ask the manufacturer or local builder/dealer for an estimate.

Financing: A log home is financed in the same manner as a conventional house. Obtain plans from the manufacturer and take those plans as well as package and construction costs to a bank for a construction loan. After the house is completed, you will need to obtain permanent financing through a bank or mortgage company. Both the VA and FHA will guarantee log homes, and secondary financing can be obtained through GNMA or FHNMA. The only real problem is in obtaining comparables or "comps." Because there may not be another log home nearby, an appraiser may have to do a bit of legwork in order to find another home of comparable construction and value.

Codes: Like panel builders, log home manufacturers ship their products around the United States and sometimes even overseas. (Log homes are popular in Japan.) As a result, their homes tend to meet or exceed any national, state, or local building codes. If there's any question, ask before you buy.

Choosing a Log Home Manufacturer: Like most factory-crafters, log home builders have a network of builder/dealers who represent them at the local level. Find out who they are in order to see any models, discuss your needs in detail, and order a log home.

The chalet style is popular among log home buyers.

Log Home Manufacturers Directory

Like panelized builders, log home manufacturers are not limited to any one specific geographical area and can ship their houses to any place in the world. Your best bet is to contact those manufacturers who interest you and get their sales literature. At the same time you can find out if they have a builder/dealer network and the location of any of their homes near you.

Manufacturers marked with an asterisk (*) grade logs under an approved grading system. For a more complete listing of the more than 1,100 log home producers and builder/dealers, you may want to get the latest *Log Home Living Annual Buyer's Guide,* published by Home Buyer Publications, Inc., PO Box 220039, Chantilly, VA 22022; (800) 826–3893.

Air-Lock Log Company, Inc. PO Box 2506, Las Vegas, NM 87701. (505) 425–8888.

*Alta Industries, Ltd., Route 30, Box 88, Halcottsville, NY 12438. (914) 586–3336.

Amerlink, PO Box 669, Brattleboro, NC 27809. (919) 977–2545.

*Appalachian Log Homes, Inc., 11312 Station West Drive, Knoxville, TN 37922. (615) 966 6440.

*Appalachian Log Structures, Inc., I–77, Exit 132, and Route 21S, PO Box 614, Ripley, WV 25271. (304) 372–6410, (800) 458–9990.

Authentic Homes Corporation, Box 1288, Laramie, WY 82070. (307) 742–3786.

B.K. Cypress Log Homes, Inc., PO Box 191, Bronson, FL 32621. (904) 486–2470.

Beaver Mountain Log Homes, Inc., RD 1, Box 103A, Hancock, NY 13783. (607) 467–2700, (800) 467–2770.

Brentwood Log Homes, 427 River Rock Boulevard, Murfreesboro, TN 37129. (615) 895–0720.

Cedar Forest Products Company, 107 West Colden Street, Polo, IL 61064. (815) 946–3994, (800) 552–9495.

Cedarlog Homes, PO Box 232, Hubbard, OH 44425. (216) 534–0182.

Cedar River Log Homes, Inc., 4244 West Saginaw Highway, Grand Ledge, MI 48837. (517) 627–3676.

Colonial Structures, Inc., PO Box 19522, Greensboro, NC 27409. (919) 668–0111.

*Garland Homes by Bitterroot Precut, Inc., 2172 Highway 93 North, PO Box 12, Victor, MT 59875. (406) 642–3095, (800) 642–3837.

*Gastineau Log Homes, Inc., Old Highway 54, Box 248, Route 2, New Bloomfield, MO 65063. (314) 896–5122, (800) 654–9253. Fax (314) 896–5510.

Greatwood Log Homes, Inc., Highway 57, PO Box 707, Elkhart Lake, WI 53020. (800) 558–5812, (800) 242–1021 (in Wisconsin). Fax (414) 876–2873.

*Green Mountain Log Homes, PO Box 428, Route 11 East, Chester, VT 05143. (802) 875–2163.

*Hearthstone, Inc., Route 2, Box 434, Dandridge, TN 37725. (615) 397–9425, (800) 247–4442. Fax (615) 397–9262.

*Heritage Log Homes, Inc., PO Box 610, Gatlinburg, TN 37738. (615) 436–9331, (800) 456–4663.

*Hiawatha Log Homes, M-28 East, PO Box 8, Munsing, MI 49862. (906) 387–3239. Fax (906) 387–3239.

*Honest Abe Log Homes, Inc., Route 1, Box 84, Moss, TN 38575. (615) 258–3648.

*Kuhns Brothers Log Homes, Inc., RD #2, Box 406 A, Lewisburg, PA 17837. (717) 568–1412.

*Lincoln Logs Ltd, Riverside Road, Chestertown, NY 12817. (518) 494–4777, (800) 833–2461. Another plant location in Auburn, California.

*Lindal Cedar Homes, Justus Division, Box 24426, Seattle, WA 98124. (206) 725–0900.

*Lodge Logs by MacGregor, Inc., 3200 Gowen Road, Boise, ID 83705. (208) 336–2450, (800) 533–2450.

Log Cabin Homes, Ltd., PO Drawer 1457, 410 North Pearl Street, Rocky Mount, NC 27802. (919) 977–7785.

Log Structures of the South, PO Box 276, Lake Monroe, FL 32747. (407) 321–5647.

*Lok-N-Logs, Inc., PO Box 613, Sherburne, NY 13460. (607) 674–4447.

*Model-Log Homes, 75777 Gallatin Road, Bozeman, MT 59715. (406) 763–4411.

Moosehead Country Log Homes, Inc., PO Box 268, Greenville Junction, ME 04442. (207) 695–3730.

Mountaineer Log Homes, Inc., Mountaineer Boulevard, Box 406, Morgantown, PA 19543. (800) 338–6346 (Pennsylvania), (800) 223–5147 (elsewhere).

*Natural Building Systems, Inc., PO Box 387, Keene, NH 03431. (603) 399–7725.

*New England Log Homes, Inc., 2301 State Street, Box 5427, Hamden, CT 06518. (203) 562–9981, (800) 243–3551. Other plants in Massachusetts, California, North Carolina.

North American Log Homes Systems & Country Kitchens, Inc., South 8680 State Road, Colden, NY 14033. (716) 941–3666, (800) 346–1521.

*North Eastern Log Homes, Inc., PO Box 46, Kenduskeag, ME 04450-6046. (207) 884–7000, (800) 624–2797. Other plants in Vermont and Kentucky.

*Northern Products Log Homes, Inc., PO Box 616, Bomark Road, Bangor, ME 04401. (207) 945–6413.

Phoenix Wood Products, Ltd., PO Box 411, Nackawic, New Brunswick, Canada E0H 1P0. (506) 575–2255.

Precision Craft Log Structures, 711 South Broadway, Meridan, ID 83642. (208) 887–1020, (800) 729–1320.

R & L Log Buildings, Inc., RD #1 Shumway Hill Road, Guilford, NY 13780. (607) 764–8275.

Rapid River Rustic, Inc. PO Box 8, Rapid River, MI 49878. (906) 474–6427.

*Real Log Homes, National Information Center, PO Box 202, Heartland, VT 05048. (800) REAL–LOG. Fax (603) 643–4321. Other plants in North Carolina, Montana, Nevada, Arizona.

*Rocky Mountain Log Homes, 1883 Highway 93 South, Hamilton, MT 59840. (406) 363–5680.

*Satterwhite Log Homes, Route 2, Box 256 A, Longview, TX 75605. (214) 663–1729, (800) 777 7288. Fax (214) 663–1721. Another plant in Colorado.

Shawnee Log Homes, Route 1 Box 123, Elliston, VA 24153. (703) 268–2243, (703) 989–5400.

*Southland Log Homes, Inc., PO Box 1668, Irmo, SC 29063. (803) 781–5100, (800) 845–3555.

*Stonemill Log Homes, 7015 Stonemill Road, Knoxville, TN 37919. (615) 693–4833.

*Tennessee Log Buildings, Inc., PO Box 865, Athens, TN 37303. (615) 745–8993, (800) 251–9218.

*Timber Log Homes, 639 Old Hartford Road, Colchester, CT 06415. (203) 537–2393, (800) 533–5906.

*Town & Country Cedar Homes, 4772 US 131 South, Petoskey, MI 49770. (616) 347–4360.

*Ward Log Homes, PO Box 72, Houlton, ME 04730. (207) 532–6531, (800) 341–1566. Another plant in South Carolina.

Wholesale Log Homes, Inc., PO Box 177, Hillsborough, NC 27278. (919) 732–9286.

Wilderness Log Homes, Inc., Route 2, Plymouth, WI 53073. (800) 852–5647 (Wisconsin), (800) 237–8564 (elsewhere).

Wisconsin Log Homes, 2390 Panaperin Road, PO Box 11005, Green Bay, Wisconsin 54307. (414) 434–3010, (800) 678–9107.

Woodland Homes, Inc., PO Box 202, Lee, MA 02138. (413) 623–5739.

*Yellowstone Log Homes, 280 North Yellowstone Road, Rigby, ID 83442. (208) 745–8108, 8109, 8110.

7

Domes— Back to the Future

Unique styles are winning homeowners over to the dome concept.

Walk into the unfinished shell of one of today's factory-crafted dome homes and you'll think you just entered the twenty-first century. For openers there are no right angles. You are, in effect, standing at the center of a hemisphere with high, vaulted ceilings and peaceful, open space (which may be the reason why one dome manufacturer sells more domes for use as churches than as family homes and commercial buildings combined).

When they were introduced in the 1960s, domes acquired the same cachet as the Volkswagen "bug": They were simple, efficient, and didn't cost a lot of money. Today, however, dome manufacturers say their structures are chosen as much for their architectural merits and reputation to withstand virtually every natural disaster except fire and floods, as they are for their celebrated individualism.

R. Buckminster Fuller, engineer and futurist, was the single most important proponent of the geodesic dome as an alternative to the traditional single-family home. He saw it as a dramatic means of creating spacious, durable, energy-efficient housing, while using a minimum amount of materials in a very economical way.

From a historic view, however, the geodesic dome owes its concept less to Fuller than it does to native building practices. Eskimo igloos, the domed huts of ancient Tunisians, the spherical yurts of nomadic Mongols—all feature hemispheres. Fuller, however, did refine the concept through his idea of constructing domes out of a network of intersecting triangles. (The triangle is the strongest geometric shape and illustrates the principle of doing more with less.)

In the United States dome homes can be found in all fifty states, with the largest number in the Southeast and Northwest. According to some estimates, the U.S. dome industry makes and sells about 1,500 kits a year, with expectations of continued growth. In addition to homes, domes are used for stadiums (Superdome in New Orleans), radar stations, theaters (Disney's EPCOT Center in Orlando), and conservatories. In Japan plans are underway for entire master-planned communities of domes. And in the Arizona desert, Biosphere II—a closed environment of connected geodesic domes—is currently housing eight environmental researchers who are fashioning a self-sustaining ecosystem.

Dome enthusiasts cite the interior openness as one of the concept's biggest pluses. And they say that the unique exterior shape helps the structures blend into natural settings. But it is their energy efficiency that draws the biggest raves.

If you compare a spherical dome with a box-type house that has the same square footage, you'll find the dome has 38 percent less exterior surface area. This means the dome requires less energy to heat or cool it. Additionally, the open interior of domes encourages better air movement and a more uniform temperature. A study done by the National Dome Council compared geodesic domes with conventional box-type houses in five climate areas. The council found that domes were 30 to 40 percent more energy efficient for heating and cooling than conventional homes. And because domes need less insulation to achieve the same energy efficiency as traditional houses, they qualify for increased financing under the Federal Home Loan Mortgage Corporation regulations.

If the idea of living in an open, well-lit, energy-efficient space appeals to you, you may be ready for a dome. Like other factory-crafted panel systems, dome kits use standard wood framing members and are available as preassembled panels or as precut components. Panelized domes are *more* labor intensive because they require a crane to set top sections into place as well as several workers to secure the panels. With component dome parts, however, an individual can erect a dome shell fairly readily by using hand tools. Either way, once the shell has been erected, the dome is finished in much the same way as other factory-crafted products.

Cost: Because domes require about a third less material than a box-type house, you can realize some savings (roughly 10 to 15 percent) right off the bat. And because a typical dome shell can be erected in under five days—and over 90 percent of dome customers elect to put up the shell themselves—you can save not only on construction time but save significantly on labor costs as well. In fact, some owners have been able to bring in a completed dome for $40,000.

Size and Style: Typically, domes are sized two ways: (1) by diameter at the hemisphere and (2) by profile at the hemisphere. Thus, you can have a 24-foot-diameter dome that is a $\frac{3}{8}$ sphere or you can have a 24-foot-diameter dome that is a $\frac{5}{8}$ sphere. The difference is that a $\frac{5}{8}$ sphere lets you install a second floor, which will add to the total square footage of your house.

Dome shells are completely self-supporting and have no interior load-bearing walls. This means you can customize a dome's interior any way you want. Most dome manufacturers also offer a variety of choices in skylights and window designs that let you decide where and how to let light in. They also provide extensions and additions you can use to make your dome house unique. If you don't see what you like or have your own ideas, manufacturers will work with you to provide custom plans and custom kits.

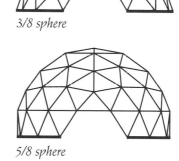

3/8 sphere

5/8 sphere

Quality: Like other factory-crafted products, dome shells are prefabricated at a common location and employ a constant labor force working within a controlled environment. Because they deal in large quantities, dome manufacturers can order a higher grade of wood than you're likely to find at the local lumberyard. This means nice, straight, almost knot-free framing material that is probably kiln-dried and therefore less likely to warp or shrink over time.

Some dome builders also include in their packages specially engineered and fabricated connectors to assure correct axial and radial angles. These connectors, which are almost impossible to get locally, make the dome structure stronger. In addition, they make the shell assembly much easier.

Time: Both panelized and precut domes will go up fairly quickly. One manufacturer says that his largest kit—a 45-foot diameter, ⅝-sphere dome—can be erected and made weather tight by three people in five days. You best bet, however, is to ask the individual manufacturer. Then factor in your skill level and whether you'll need a crane to help set panels (if it's that type of package). Because they still make up only a small fraction of the home industry, most dome manufacturers do not have a builder/dealer network and rely on customers to spread the news about them. But the construction process used to erect the shell is fairly straightforward, and most experienced builders should have no trouble putting one up.

Codes: Because dome manufacturers operate all over the United States—and in many cases overseas—their plans meet or exceed any national, state, or local building codes. If your local building inspector has any problems, a quick call to the company should solve them.

Choosing a Dome Manufacturer: Since dome factory-crafters are relatively few in number, it will probably be easier to select one than it would be with a factory-crafter that has a larger universe. Bear in mind the maxim of comparing apples with apples. Find out what each manufacturer includes in its package as standard. Some feature special fasteners that make it easier to assemble the triangles that make up the completed dome. As mentioned earlier, there are few builder/ dealer networks. Most manufacturers should be able, however, to provide you with a list of customers with homes fairly nearby that you can visit. And since dome owners are enthusiastic about their style of living, you should be able to get some good reports as well.

Spacious interiors allow natural sunlight to fill a dome interior.

Dome Manufacturers Directory

Like panelized and log home builders, dome manufacturers can and do distribute their products worldwide. You will probably want to call or write those who interest you and get their sales literature. At the same time you can also find out if they have either a model or spec home nearby that you can check out.

Domes America, Inc., 6345 West Joliet Road, Countryside, IL 60525. (708) 579–9400.
Geodesic Domes & Homes, 608 Highway 110 North, PO Box 575, Whitehouse, TX 75791. (214) 839–7229.
Geodesic Domes, Inc., 10290 Davison Road, Davison, MI 48423. (313) 653–2383.
Oregon Dome, Inc., 3215 Meadow Lane, Eugene, OR 97402. (503) 689–3443.
Timberline Geodesics Inc., 2015 Blake Street, Berkeley, CA 94704. (415) 849–4481.

8

Selecting House Plans

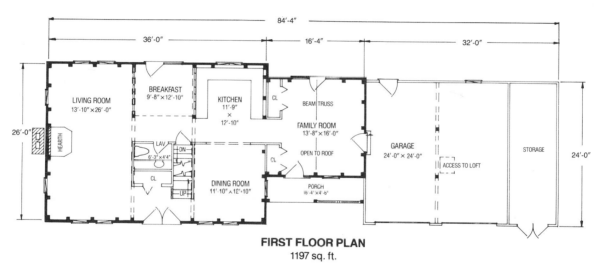

FIRST FLOOR PLAN
1197 sq. ft.

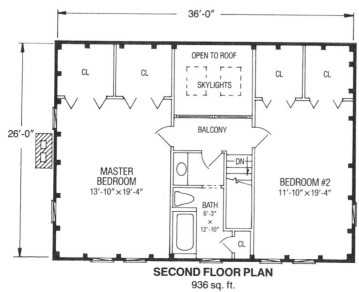

SECOND FLOOR PLAN
936 sq. ft.

The traditional post-and-beam concept is enjoying a comeback in the panelized realm.

Most factory-crafters offer a good array of standard styles and plans that should please anyone's taste and budget. If not, most are willing to modify a standard plan to your specific needs. And if you still can't get what you want, most have an in-house design staff that can work with you, generally for less than you'd pay an architect or home designer for conventional house plans.

But whether you opt for standard or customized plans, bear in mind the following rules of thumb:

- Every cut you make in size will save you money.
- One-story homes are great, but it's cheaper to build the same square footage in two stories or with basements.
- A simple shape is easier and cheaper to erect than one with a complex foundation, roof, and interior ceilings.

Looking at Plans

You may be able to get a good preliminary idea of what plans fit your needs by looking at manufacturers' promotional literature. But once you're serious about a particular house, you will ultimately need to contact the factory-crafter and get a full set of plans. They customarily cost a few hundred dollars, which will be subtracted from your package cost once you go ahead and place your order. These plans will then be used by contractors for bid estimates and by the bank and appraiser for evaluating your loan application.

You will generally find factory-crafters and their builder/dealers very helpful in assisting you to find the best plans for your site and your needs. Often they can assist you in finding a lot as well. If you already have one, they can aid you in suggesting a particular plan or the best way to place your chosen plan on your lot. At this time they probably can also give you a ballpark estimate on how much it will cost you to buy the package and finish the house (normally stated on a cost-per-square-foot basis). Normally this is 2½ to 3½ times the price of the kit, not including land. This ballpark figure varies, however, according to what is included in the kit's components, the local labor market, and other material costs. Ask the manufacturer what its general formula is. Bear in mind that the more complete the factory-crafted house package is (like modular), the more accurate the estimate will be.

If you're not used to reading house plans, think of them as a detailed set of visual and written directions from an architect or house designer to contractors on how to build your house. You don't have to know all the specialized terms, and you don't have to understand all the detailed drawings. But you should be somewhat familiar with what they include:

- Foundation plan
- Floor plans for each level
- Exterior elevations for all four sides
- A detailed sheet or book
- Specifications

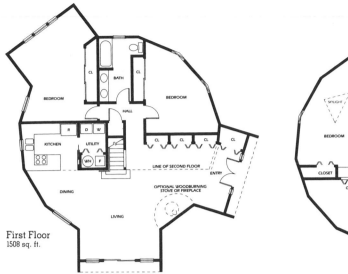

First Floor
1508 sq. ft.

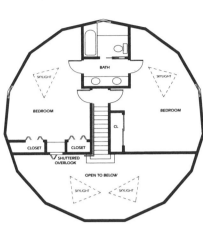

A big advantage of dome structures is their ability to enclose a lot of space with less material than a traditional house. In these plans, almost 2,200 square feet are enclosed by a 40-foot diameter sphere. If this were a stick-built house, it would have to be 40 by 50 feet.

Foundation plans should show the footprint of the structure from the footings, the part of the foundation that rests directly on the soil, up to the sill plate, wooden members that sit directly on top of the foundation wall. Homes are built on a slab, crawl space, or basement foundation, and your plans should reflect which kind will be used. The type depends on the topography of your land. You need not be too concerned with foundation plans other than to make sure they accurately describe the type of foundation (slab, crawl, basement) and show a correct outline or footprint of your home.

Floor plans should show the position of all interior elements, including walls, cabinets, doors, windows, fireplace and hearth, tubs, showers, lavatories, commodes, sinks, plus any load-bearing posts and beams. I have found it helpful to walk around inside a house mentally via the floor plans. I try to picture each space in my mind and see if I'm comfortable with the way I can move from room to room. Additionally, all plans are dimensioned. I like to take a 25-foot steel tape (available from any hardware store) to the building site and mark off spaces so I can get an idea of how big or small a space is. For example, your master bedroom might be 15-by-20 feet. Mark this off. Walk around in it. Are you comfortable with this space? Is it too small? Where will your furniture fit? If you want to, can you enlarge the space without a lot of expense? Notice the way interior doors swing (to the left or right, into a room or out toward you). Pay attention to closet space. Is it enough? If not, can you add more without a major redesign?

Exterior elevations should show window and door positions and details, siding and/or brick, roof details, skylights, and chimneys. In other words, all the details of what your house will look like from the outside. Does it look like you thought it would? Are there any additions or deletions you want to make? Ask yourself if you want more windows. Usually they're easy to add and won't cost you too much more.

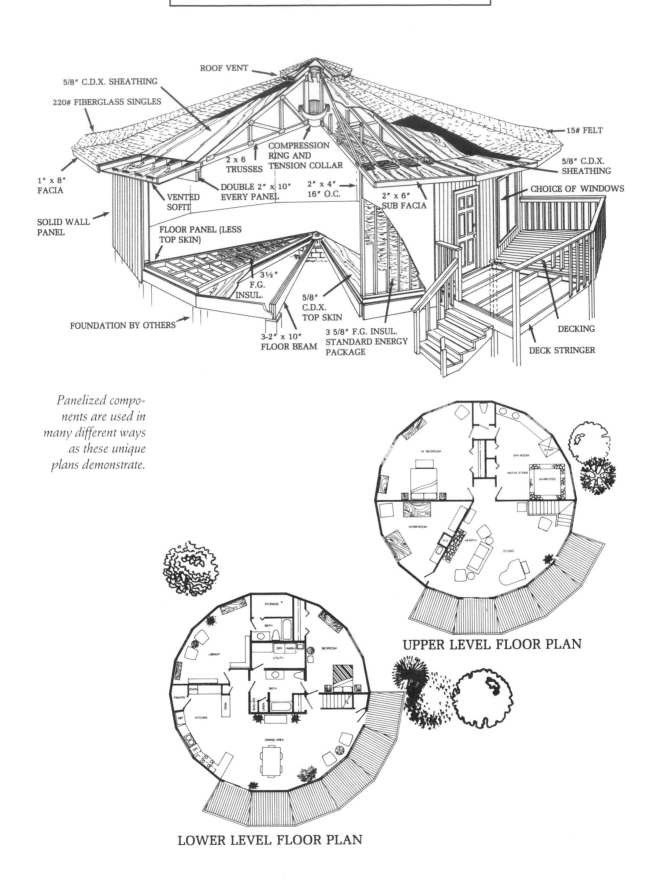

ROOF VENT

5/8" C.D.X. SHEATHING

220# FIBERGLASS SINGLES

15# FELT

5/8" C.D.X. SHEATHING

COMPRESSION RING AND TENSION COLLAR

2 x 6 TRUSSES

1" x 8" FACIA

DOUBLE 2" x 10" EVERY PANEL

2" x 4" 16" O.C.

2" x 6" SUB FACIA

CHOICE OF WINDOWS

VENTED SOFIT

SOLID WALL PANEL

FLOOR PANEL (LESS TOP SKIN)

3½" F.G. INSUL.

5/8" C.D.X. TOP SKIN

DECKING

FOUNDATION BY OTHERS

3-2" x 10" FLOOR BEAM

3 5/8" F.G. INSUL. STANDARD ENERGY PACKAGE

DECK STRINGER

Panelized components are used in many different ways as these unique plans demonstrate.

UPPER LEVEL FLOOR PLAN

LOWER LEVEL FLOOR PLAN

Detail book/sheet(s) should show wall cross-sections, framing details, and anything that may not be clear from the floor plans. This is pretty much contractor-oriented, and you shouldn't have to get involved with this level of detail.

Specifications. Most manufacturers include a list of factory specifications and factory options with their promotional material (Andersen windows, Peachtree doors, ½" CDX fir plywood sheathing). These specs give you some basis of comparison when you're looking at different companies within the same category. As with the car industry, what's factory-standard and what's factory-option varies from manufacturer to manufacturer. Usually the more complex the finishing process at the job site, the more chance for input you have in the specifications. Modular and mobile homes leave the factory fairly complete, and you only get to choose such things as wallpaper, carpet color, and siding. Panelized, log, and dome homes leave the factory in pieces (small and large), and you often get to choose from a greater variety of options.

As part of your house plans, specifications list everything that will be going into your home, from what kind of floor joists will be used to what kind of plumbing and fixtures you want. They help you, your lender, your suppliers, and your contractor/subcontractors know exactly what is going into your house. They are necessary so that you can control cost and quality. Professional builders need them in order to give accurate bids. You need them so that you can know what to budget for and how much to budget. Most factory-crafters have complete specifications for the homes they sell, and they or their local builder/dealer will be able to help you make selections based on your pocketbook.

For some items—floor coverings, cabinets, appliances, wallpaper—builders normally establish an "allowance." This means that, while you may not have made a selection right now, you will need to "allow" for a certain sum to be spent at a later date. Just think of it as budgeting for future expenses so that you can complete your total cost estimate. Suppliers, your builder/dealer, or the factory-crafter can help you establish reasonable allowance figures provided you tell them the level of finish that you want (basic, medium, deluxe) or the amount of money you have available to spend.

Making Changes Early

I like to tell people that it's much easier (and cheaper!) to make changes on paper than it is at the job site. For that reason, make sure you're comfortable with the plans and specifications. If there's something you don't like, talk with your builder/dealer or the manufacturer. See if they can suggest a change. If you reach an impasse, maybe you've got the wrong plans or the wrong house. It's better to back out now than it is to start building.

9

Buying the Land

The three most important things to consider in buying land are location, location, and location.
— Old real estate saying

Regardless of what type of factory-crafted home you plan to buy, you will have to have some land on which to put it. If you've already been in touch with a builder/dealer for a particular factory-crafter, that person may know of some choice lots around your area on which to build. Builders tend to keep a visual inventory of potential sites and are normally willing to help you select a site that fits your budget and your plans.

While not a difficult process, buying land is a very personal decision, and there are certain guidelines to follow that can help you. Two of the most important are cost and location.

Land Costs

Land costs can vary more than any item in the building business. Where I live (Charlotte, North Carolina), you can buy four acres in the country for $20,000. But a smallish lot in the city could run as high as $300,000. It's really the old law of supply and demand in action. Heavy supply, low cost. Heavy demand, high cost.

How much you should spend on land depends on your needs and your pocketbook. A good rule of thumb is to spend no more than 15 to 20 percent of the appraised or market value of your finished house. This means that if your house will cost you $80,000 (package price plus construction costs plus appliances and so on), you should probably pay no more than $16,000 for your lot. But there are always exceptions. One individual I know paid $80,000 for a lot on which he planned to erect a house costing $185,000. Why? He liked the location.

Bear in mind that if you have to put in a well for water and/or a septic system for a sewer, you will likely add about $5,000 or more to the cost of your house.

Location

Since houses are site-bound, it's a good idea to think hard about your location. Topography, tree density, soil conditions, availability of water and sewer connections, and accessibility are some of the general factors you will want to consider as you look for a building site.

If you find a sloping lot, you may be able to build a basement foundation. This will increase the amount of useable square footage of your house without adding a lot of extra costs. Additionally, sloping lots assure good drainage no matter what type of soil. They allow for one side or more of the basement to have good window areas for light and dryness. And, depending on the slope, one open side can be frame construction, which is less costly than foundation block or concrete.

If you like trees, try to find a wooded lot on which to build. Bare lots are easier to build on but cost more to landscape. A wooded lot generally costs more to buy but less to landscape. And a wooded lot is more appealing when it comes to future resale as well; people generally prefer lots with mature trees.

Using a Real Estate Agent/Broker

Most real estate agents and brokers are experienced professionals who can help you locate the property you want and guide you through all the complexities inherent in making a land purchase. Many are tied into computer networks that can spit out in a matter of seconds information on available lots that could take you weeks to uncover on your own.

Strike up a friendship with a local agent or broker. Show him or her what kind of house you want to build, what kind of a location you think you want, and what price you'd like to pay. Remember that the *seller* pays their commission, not you.

You can find agents and brokers in your local yellow pages, through the recommendations of friends, or just by driving around an area and looking for signs. Many agents are members of local chambers of commerce. Other sources include the local association of real estate agents or the National Association of Real Estate Brokers (NAREB).

Some agents may not be sold on the idea of factory-crafted homes. Show them this book and the promotional literature from the manufacturer. Have them talk with the local builder/dealer. Take them to see a model. If necessary, change brokers. As a final caveat, keep your own needs uppermost in your mind and let the agent help you but not force you into a decision. You are the best authority on what you want.

This woodsy lot suits this new modular home perfectly.

A Land-Buying Checklist

Other factors to consider when buying land include:

Neighborhood

☐ *Will your house fit in with other homes in terms of style, size, and cost?* This can be important for resale later.

☐ *Are there good schools, churches, and shopping facilities nearby?*

Transportation

☐ *How long will it take you to get to work?*

☐ *Will you be able to reach major roads fairly easily?*

☐ *Are major roads planned for your location? Where? When?*

☐ *Is public transportation accessible?*

Pollution

☐ *Are the levels of air, water, and noise pollution acceptable?*

Utilities

☐ *Is community water supply, sewer, and waste removal available?*

☐ *Are other utilities available (phone, electric, cable TV, gas)?*

Physical Characteristics

☐ *What (if any) is the lot's direction of slope? Toward the proposed building site? Away from it?*

☐ *What are the building characteristics of the soil? Is it sandy? Rocky? Claylike? Will it support concrete footings and a building without a great deal of excavation?* (You may have to consult with an engineer on this to find out.)

☐ *If you have to install a septic system, will the soil percolate?*

Physical Characteristics (continued)

☐ *Is there a flood plain nearby? How close is it to your proposed foundation?* (Probably your local authorities won't let you build too close, but a flood plain can affect things like the location of a swimming pool.)

☐ *Is the lot big enough or private enough for your purposes?*

☐ *What amount of clearing will have to be done for the kind of house you want to build?* Tree removal prices vary from area to area. Small trees can be taken down by your grading contractor, big trees by a tree expert.

Zoning and Restrictions

☐ *How is the area zoned? Is it strictly residential or is it a mix?* To find out if it may change in the future, check with your local planning officials.

☐ *What are the deed restrictions about building size and set backs?* These define the building "envelope" and may limit your home's position on the lot.

☐ *Is there a neighborhood design review process?* A real estate agent will know this and tell you who to contact. Usually such review processes are a formality.

Deed

☐ *Is there an unencumbered deed?* If not, pass on the lot.

☐ *Does the deed include easements (utility, access)? What kinds? Where?* Utility easements are pretty normal and should not infringe on your plans. Access easements may affect where you want to place your home.

One final word on building a factory-crafted product in an existing community: Panelized and precut homes tend to stir up the least resistance because they most resemble the production system by which site-built homes are done—they require the most amount of site work. Modular homes tend to stir up greater resistance because they arrive at the job site on a trailer (or trailers) and thus fall outside the traditional view of how a home should be built. Log homes and domes are generally built in rural areas or on large parcels of land, so the question of community resistance usually doesn't apply. Mobile homes meet the greatest resistance because of a lingering public perception of them as unsafe and unconventional. If you do meet resistance, find out if it's backed up by legal restrictions. If so, you can either pass or fight. If there are no legal restrictions but your neighbors are balky just the same, try giving them a quick educational program showing that your factory-crafted home will be the equal (or better) of anything in the neighborhood.

10

Getting the Money

To finance your factory-crafted house, you'll need to secure two kinds of loans: a temporary construction loan and a permanent loan or mortgage.

Construction Loan
- Higher interest
- Short term (months)
- Pays building costs

Mortgage
- Lower interest
- Long term (years)
- Pays for total

The Construction Loan

A construction loan allows you to draw money in stages that relate to the completion of your house. It's like having a checking account to pay for building your house, only you need the bank's approval—via its construction loan officer, who periodically inspects the job site to see what progress has been made—to write the checks. You may have to pay some expenses before receiving your first draw, but they usually don't amount to more than a few thousand dollars. You will get the money for these expenses paid back to you when you get the next draw. If you don't have the funds to cover these initial expenses, you could borrow enough from a commercial bank to cover them. Such a loan shouldn't be hard to get because the source of repayment will be your construction loan.

The Mortgage

A mortgage comes into play once the house is built. A mortgage is a loan that has the repayment stretched out for a long period of time, usually fifteen or thirty years. Most lenders require that you quality for a permanent mortgage *before* they lend you the money to build. This assures them that there will be a "takeout" lender at the end of the building process (and they won't be left holding the bag should you default). Quite a few lenders prefer to do both the construction and the permanent loan. If you're going to use one lender for construction financing and another for the permanent mortgage, you will need a letter of commitment from the permanent lender to show the construction lender.

Getting Started

To start the process, shop the local financial marketplace. Your best bet is to find a lender that will do *both* types of loans. If you can't, find out if you can qualify for a permanent mortgage first, then search for a lender for the construction loan. If you're using a builder/dealer, he or she will probably have contacts and can help. Insofar as loans are concerned, banks see themselves as being in the risk-taking business and therefore look at your personal financial information to see (a) if you have the means to repay the loan and (b) if you have a good history of repaying others.

You don't need to have your house plans in hand when you're shopping around for a mortgage. You should, however, have some rough idea as to the total amount of money you need to borrow (land plus kit plus construction costs). The manufacturer or builder/dealer should be able to give you these approximate figures. When you present them to the bank, err on the high side since it's easier to cut your loan than it is to increase it.

Assuming you can qualify for a permanent loan, you can take the next step and apply for a construction loan. For this you will need a complete set of plans and specifications, plus an estimate from a general contractor stating what it will cost to build your house. Most states will let you build your *own* house without being a licensed general contractor. But banks see the situation differently. Chances are they will require that you hire a general contractor or give very convincing evidence that you are indeed qualified to be your own general contractor. They want to make sure the house gets built correctly and within budget, so if a default should occur, their investment will be protected.

One of the agreeable side benefits of using the factory-crafted system is that you help alleviate a bank's fears of cost overruns or failure to complete the house. If you're building a modular house, for example, you can assure the bank that the house will be delivered to your lot about 95 percent complete and will require only a few draws for its completion:

Dubbed the "Modular Mansion," This 5,700-square-foot home in Mamaroneck, New York, was designed and built to show the flexibility and range of modular building

foundation, package COD, "buttoning-up" (connecting utilities, minor carpentry and painting), cleanup, and landscaping. If you're building a panelized, log, or dome structure, you can again assure the bank that all the house components are largely included in the package and that your cost and time estimates for finishing the house are based on the manufacturer's proven experience.

Before You Go To the Bank

Applying for a construction loan is basically a matter of having all your ducks in a row. In this case your "ducks" include:

- Owning the land on which you plan to erect your factory-crafted house. If you don't own it, try to have it under contract for purchase with the contingency that you can get financing for it. If neither is possible, at least have your property selected.
- Having a complete set of house plans from your factory-crafter, a survey of your property (with surveyor's seal), and a copy of your deed with its restrictions.
- Having an accurate estimate of all your building costs. If you're using a builder, you'll have to bring that person's figures with you. If you're acting as your own builder, you'll have to have all your major subs and suppliers lined up.

What you're doing is essentially a sales job. You're pitching your idea for a factory-crafted house to someone you hope will lend you the money. Here are some tips on making an effective presentation.

1. Be concise. Experts urge that you get to the point within thirty seconds. "Customers want to know what time it is, not how you built the watch," says Andrew Gilman of Comm-Core, Inc.

2. Practice. Make sure you know what you're going to say and in what general order.

3. Dress conservatively. This is especially true when meeting bankers.

4. Bring visuals. If you're selling a factory-crafted home, a lender will want to know what the finished product will look like. Many people misunderstand the factory-crafted concept.

5. Answer directly. If you don't know the answer to a question, promise to find out and then follow through.

6. Relax. Be yourself. Lenders are ultimately buying you as well as your ability to repay.

7. Persist. If one lender says no, try another. A right idea always can be fulfilled.

11

Estimating Costs

Estimating the cost of building your factory-crafted home is a vital part of the financing process. As mentioned earlier, when you begin the loan process, you need to have an idea first of your rough costs and then a more accurate estimate of your construction costs.

Most manufacturers have a rule of thumb they use to provide a rough estimate of the cost of their finished homes. Modular manufacturers normally base theirs on the square footage of the particular home that has been selected. Mobile home manufacturers usually base theirs on the model selected and where it will be placed. Panelized, log, and dome manufacturers give either figures that are 2½ to 3½ times the kit price or the average cost of construction per square foot for a particular area of the country.

The only way to get a more detailed estimate is to list all the individual labor and material items that go into making up your house and get estimates on as many of them as you can. If you plan to use a general contractor (and the lender may require that you to so do), that person will do the estimating and add in his or her profit and overhead. If you plan to act as your own general contractor (lender permitting), you will have to gather the costs yourself.

The following is a typical list of items that go into finishing a panelized, log, or dome home. Modular and mobile homes—which leave the factory almost totally complete—are considerably less complicated.

Sample Cost List

Here is an honest-to-goodness sample of items from a panelized house (with walk-out basement) that I built. The manufacturer's package was very complete and included all materials needed to make the structure weather tight. It also included deck materials, all interior trim, and all interior and exterior doors, windows, and hardware. If you're building a modular home, only those items followed by an asterisk (*) will not be included in the package. Mobile homes are a special category. Cost estimates for them can best be supplied by a local dealer.

This list, of course, was for a specific house in a specific location. Your list may be different. As mentioned earlier, if you're building a modular home, you won't have to deal with many of the items because your house leaves the factory almost complete. If you're building a log home, the log walls themselves provide good insulation, and you may want to leave the natural wood rather than apply wallboard. Domes don't require gutters and downspouts. And so on.

Your lender will require you to get fairly accurate costs on completing the house *before* you begin building. In theory, this sounds easy. But many contractors are unwilling to give you a firm price without seeing exactly what they're getting into. For example, drywall contractors can look at your plans and give you a rough estimate of what it will cost to install and finish drywall in your house. But until they can actually walk around inside the structure, they will be unable to give you an exact figure unless they have done an identical house before. In these cases your best bet is to ask the manufacturer what the costs have been in the past with house plans identical or similar to yours. Then translate that figure into your local situation.

As you start, you will probably find that some costs go over what you estimated. If so, you can look for ways to save by using less-expensive materials or by doing without a certain item until later (like a garage, for example). How to make sure labor costs do not inflate is discussed in chapter 13.

The solarium area inside an Acorn Country House.

Here is the list:

1. Land (*). You should have a firm price on this.

2. Survey (*). You will need to budget for several surveys. (1) When you apply for your loan, your lender will require a survey of the property to accurately show the boundaries of your land. It must be done by a registered surveyor and stamped with that person's seal. (2) You will need another survey to get your land staked for clearing. (3) You will need to have your surveyor stake where the footings should go. (4) After these are in you should have the surveyor back to set corner pins or batter boards so that the foundation crew can install an accurate foundation. *This is vital. Most factory-crafted homes are made to fit a specific-sized foundation. Check with your manufacturer for acceptable tolerances.* (5) Your lender will require a building survey at the end to show that the house was built per plan.

3. Plans and Specifications (*). You will already have paid for these from the manufacturer.

4. Closing Costs (*). Your lender can detail these for you. They normally include a service charge, points, attorney's fees, title insurance, taxes, recording fees, and any other fees the lender may charge. Often you can save on closing costs by using the same lender for both the construction loan and the permanent mortgage.

5. Builder's Risk Insurance (*). Your lender will require that you carry insurance on your home while it is under construction. It is necessary in the event of fire, damage, or theft. When general contractors are used, the insurance is their responsibility, and they will add it to their costs. The price of insurance varies with the insurance company. It does not cover people, and you may want to look into a general liability policy to cover anyone other than subcontractors who may be injured on the job site. General contractors normally have a general liability policy for themselves and require all their subs to carry their own liability and workmen's compensation.

6. Construction Loan Interest (*). This can be estimated by your lender. It will vary according to the type of factory-crafted house you intend to build and the location in which you are building.

7. Building Permit (*). Permit fees provide for on-site inspections by your local building department. They vary from locality to locality. When general contractors are used, they will have to obtain the permit.

8. Temporary Electrical Service. The electrician usually provides a temporary electrical outlet so that your local utility can set a temporary meter. This is so that all subcontractors who require electrical power during the construction of your house can have it. The cost for this is nominal. Once your house is wired and power run to it by the local utility, a meter is set, and you can use the outlets in your home.

9. Portable Toilets. Portable toilet companies are listed in the yellow pages and charge a monthly rental fee. Multiply this figure by the time you think it will take you to build.

10. Site Preparation (*). Your grading subcontractor can give you a price to clear and rough grade your lot after having a chance to see and find out what you want to build and where you want to build it.

11. Excavation. If you are building a house with a basement, your grader can also give you a price on what it will cost to dig out for the basement.

12. Fill and Gravel and 13. Driveway Fill. Different sites have different needs. If you have a sloping lot you may need to bring in some extra fill to raise the grade in places. A good grading contractor should be able to give you an estimate.

14. Footings (*). A footing subcontractor should be able to give you a price (usually stated in linear feet) for footings, including concrete and labor. If access to your job site is difficult, then a pump truck will be needed to pump concrete to the footing excavation. This will cost additional money.

15. Foundation Materials and Labor (*). A brick-and-block mason should be able to give you a fairly accurate estimate on materials and labor for your foundation. As an alternative you may want to consider a concrete wall foundation. It generally costs more but is stronger and does a better job of resisting water. Wood foundations—which use pressure-treated material to resist water—are gaining popularity in some areas. Like concrete, they cost more to install, but users claim they make for warmer basements.

16. Waterproofing. If you are building a house with a basement, you will need to get a price from a waterproofing subcontractor. Do not try to save money here. Go with the one who can give you the best *guaranteed* job.

17. Termite Treatment (*). Some areas of the country require that the soil be treated before major construction begins. Many lenders require an additional treatment once the house is completed.

18. Inside/Outside Drains. You only need to consider these if you have a basement house. The waterproofing subcontractor (see item 16) should be able to give you a price.

19. Basement Slab. A concrete subcontractor should be able to give you an accurate estimate (quoted as a per-square-foot price for materials and labor). Make sure the estimate includes any necessary fill for preparation, Styrofoam insulation, plastic film, and reinforcing steel as required by plan or code.

20. Factory-crafted Package. This price is from your manufacturer. Make sure you know

what it includes so you will know what, if any, additional materials will be needed. Ask if freight is included or if you will have to pay for it on delivery.

21. Crane (*). When the package arrives at the job site, you will have to unload it and place it near the foundation. Sometimes a forklift can do this. Other times you will have to rent a crane. The yellow pages list companies that provide these services. Rates are generally figured on an hourly basis.

22. Framing (Package Erection). You can get a quote for this from a carpentry subcontractor. It will probably be based on the square footage of your house. The price will vary from location to location but should include all labor for making your house weather tight and for building any decks. Sometimes one subcontractor can handle *all* carpentry items including cabinets, shelving, extra built-ins, and interior trim. Where I live, however, usually one crew does the rough work (framing) and another the trim.

23. Roofing Labor. Roofing subcontractors usually charge by the square (100 square feet = 1 square) to fasten shingles, and they charge by the linear foot to do flashing. Usually your factory-crafter can tell you how many squares your house will have and how much flashing you'll need. The roofer's price should also include installing ridge vents (if used) and capping. Warning: Do not always go with the lowest estimate. A good roof (and a good roofer) are worth their weight in gold, as you will find out in a heavy rainstorm.

Log homes can feature stunning, open interiors.

24. Deck Labor. See item 22.

25. Heating and Air-Conditioning (*). A qualified subcontractor should be able to give you a firm price based on your specifications and the specifications provided by the manufacturer. Be sure the price includes any venting for bath fans, stove or cooktop vents, furnace venting for gas or oil, dryer vent(s), and any other required venting.

26. Plumbing and Fixtures. Plumbing contractors provide all the labor and materials to plumb your house, including all fixtures except the dishwasher, disposal, and washing machine. You will want to make sure that you get a price to hook up these items as well as any ice-making attachment your refrigerator may have. Your fixture specifications to the plumber must be very clear. In some cases a plumber may be able to suggest a less-expensive alternative, should you be running over budget. The plumber's contract price should also include any costs to connect water and sewer lines to their source or to install any needed septic tank, pump, or system. If your property requires that you dig a well, you will have to factor this cost in early on.

27. Electrical and Fixtures. An electrical subcontractor should be able to give you a firm quote on wiring your house according to the local building-code requirements and installing your fixtures. The price should include the rough and trim work needed to supply and install all switches, receptacles, panel boxes, circuit breakers, any exterior lights, security systems, intercom, and built-in stereo system. It should also include wiring any built-in appliances and/or heating and air-conditioning equipment.

28. Chimney and Fireplace (*). Fireplaces are expensive, but most homeowners want them. If you want to save money, install a nice wood-burning stove instead. If you're convinced that a fireplace is necessary, your brickmason should be able to give you an estimate based on time and materials. If the mason doesn't supply the materials, your masonry supplier should be able to give you an estimate on amounts. In either case, treat the cost as an *estimate* and realize that one of you will be a little off.

29. Insulation Materials and Labor. If you're building a closed-wall panelized house, you may not need to figure this expense. Otherwise, an insulation subcontractor should be able to give you a firm price on materials and labor to insulate your house per manufacturer's specifications.

30. Gypsum Wallboard Materials and Labor. This is commonly called *sheetrock* by the trade. Usually one subcontractor handles all phases of this operation: stocking, hanging, and finishing. The price may be quoted to you as a specific amount or as a per-board price. Normally an experienced sub can accurately tell you how many boards it will take to hang a house. Make sure this price includes any necessary scaffolding, taping, finishing,

and scrapping (getting rid of leftover material). The subcontractor should also come back just before the painters and "point-up" the house (fill any small dings or cracks).

31. Trim Carpentry Labor. See item 22. If your framing carpenter doesn't do trim work, you will need a separate price from a trim carpenter. Again this is given on a square-foot basis and should be lower than the framing price (by at least a half or two-thirds). Make sure the price includes all interior doors and installing all trim such as stairs, handrails, window trim, baseboards, and door hardware.

32. Painting and Staining Materials and Labor. You can obtain a quote—sometimes given on a square-foot basis, sometimes given on a contract-price basis—for all exterior and interior painting and staining. *Never pay on the basis of time and materials.* Make sure the painter knows what you want done and what materials you want to use. Often a painter can suggest these if you're unsure. To save money, you may want to spray the exterior (except trim) and the interior prime coat. Check to see if your painting contractor can do this and how much you can save.

33. Ceramic Tile. A tile subcontractor can give you a firm price based on the materials you want to use and the areas you want tiled. Find out if such accessories as towel bars, soap dishes, and marble thresholds are included in the price.

Amenities like this whirlpool can make your dream house truly dreamy.

34. Shelving Materials and Labor. If closet and kitchen pantry shelving doesn't come with your factory-crafted package or if it does come and your trim carpenter will not install it, you can get a firm price from a shelving subcontractor based on what you want and where you want it.

35. Kitchen Cabinets/Countertops. Included in this category are bath vanities as well as kitchen cabinets. They can be obtained from a building supply company or a firm that specializes in kitchens and baths. Depending on your tastes and pocketbook, the cost for these can be sizeable.

36. Gutters and Downspouts Materials and Labor (*). Some homeowners feel they can do without these. My preference is to get water *away* from the house whenever I can. You can obtain an estimate from a gutter subcontractor based on the linear footage required.

37. Garage Door Materials and Labor (*). You can get a firm price to supply and install a garage door or doors (and remote-controlled door opener) from a subcontractor who specializes in this. If you shop around, you can usually save money.

38. Driveway, Walks, Patios (*). Using a copy of your plot plan, which shows where your house will be located on your lot, a concrete subcontractor should be able to give you a good estimate. If you want to use other materials such as crushed stone or brick, you will need to contact the appropriate subcontractor.

39. Garage Slab (*). See item 19.

40. Cleanup and Disposal (*). You will periodically have to have the job site cleaned up and the resulting trash disposed of. You may want to rent a small dumpster and have it emptied on a regular basis. Then, once your house is completed, you will need to have everything inside the house cleaned, including windows, bathrooms, cabinets, and closets. Companies listed in the yellow pages who specialize in this can give you an accurate price.

41. Finish Grading (*). Normally your grading contractor will be able to estimate what this will cost when figuring your site preparation (item #10).

42. Hardwood Flooring Materials and Labor. Generally one subcontractor can install the flooring and finish it. Prices are typically quoted on a square-foot basis. If you're buying the materials, make sure you plan on buying about 30 percent more hardwood than your plans require to account for matching and waste.

43. Landscaping (*). Based on your plot plan, a good landscaping contractor should be able to provide you with a price to supply and install the shrubbery, trees, and other landscaping you want.

44. Appliances. Appliances to include in your estimate include stove or oven, range or cooktop, garbage disposal, refrigerator, dishwasher, microwave, trash compactor, grill, and so on. Price shopping here is always a good idea.

45. Carpeting. A floor covering supplier should be able to give you an accurate price based on your plans and what grade carpet and padding you want.

46. Miscellaneous Costs and Overruns (*). No one can expect to account for all the costs that go into building a house. Therefore it's best to add up items 1 through 46 and multiply by 5 percent. This will give you a contingency figure to dip into as the need arises. And now is the time to add any items not covered in the list, such as swimming pool, additional decks, mailbox, and so on.

12

Building Your
Factory-Crafted Home

A custom-built modular home.

There are two ways to realize the dream of owning your own factory-crafted home: (1) hire a general contractor (builder) or (2) do it yourself (see chapter 13).

Building a new home is a team business. Your team should include your real estate agent, loan officer, insurance agent, lawyer, accountant, factory-crafted home manufacturer, and general contractor. Hiring a contractor to build your house can be a fun and rewarding experience. It's really the first concrete step to realizing your dream of a new home.

Contract Fees

In general most contractors are professional, law-abiding citizens who bring managerial skill and building know-how to the home-building business. Most work on a specified contract-price basis. This means they will estimate how much it will cost them to build your house (less the package price) and add onto that a fee for managing the project. Normally this fee runs around 10 to 20 percent of the total cost, but with complex houses it can run higher.

Some builders will work on a cost-plus-fee basis. This means they will build your house for whatever the costs run *plus* their fee. The problem with this method is that it puts the risk entirely on the homeowner's shoulders. The builder is assuming little risk and therefore has little incentive to hold costs down.

When you begin to solicit bids on your house—either from the manufacturer's local builder/dealer or from other contractors in your area—try to picture your negotiations as a "win-win" situation. You want to win by getting the best house for a reasonable price. Contractors want to win by building the best house for a reasonable price. After all, they want your good reference for their advertising.

Follow the same rules you would in spending any large sum of money. Get about three independent bids from as many contractors and check homes they've built and their references. Then go with the individual you think will do the best job and with whom you feel most comfortable.

Don't always go with the builder who gives you the lowest price. Experienced professionals know how to cover contingencies when they estimate a job and will most likely be able to do a good job for the price they quote. If a bid is too low, you might end up with a job that has not been estimated correctly. This could result in a builder/homeowner relationship that gets strained. Again, my advice is to go with the individual with whom you feel most comfortable.

A.

B.

C.

A. Factory workers use an over-head crane to position a completed roof section on a module for a one-story home.

B. Painting and roofing work occur simultaneously as part of the efficiency made possible with factory-regulated construction.

C. Modular components, 95 percent complete, are shipped from the factory to the home site where they will be assembled.

Finding a General Contractor

If your factory-crafter does have a builder/dealer in your area, you don't necessarily have to use that individual to *build* your house. But you will most likely have to go through that person to *buy* your house. Chances are you cannot buy directly from the manufacturer any more than you can buy plumbing fixtures or appliances directly from the manufacturer. But you may want to consider using an established builder/dealer if there is one in your area. Such an individual may have a model that you can inspect, will know which options are useful for you, will probably have built a few of that manufacturer's products, and will probably use subcontractors that are familiar with the type of house you want to erect.

If you don't want to use the established builder/dealer or if the manufacturer has no established builder/dealer in your area, you can find a residential contractor by:

- Looking in the yellow pages under "Building Contractors" or "Home Builders." But also realize that sometimes the best ones don't have a listing here.
- Writing to the National Association of Home Builders (NAHB), Fifteenth and M Streets, NW, Washington, DC 20005, for a list of NAHB members in your area.
- Seeing if your state requires contractors to be licensed. If so, there should be a published directory of all licensed builders.
- Checking with your local building inspectors. They have a notion of who's good and who's not.
- Asking your real estate agent for recommendations.
- Checking with your local chamber of commerce, local home builder's association, or bank mortgage department.
- Driving around to places where new building is going on. Look for quality work and find out who the builder is.

Today's general contractor is likely to be more of a manager than an actual builder. As someone recently put it, "Fully 90 percent of my time is spent solving problems . . . getting people to work together . . . the rest of the time is spent building."

Contracts

When you go to apply for your loan, you will want to bring a signed contract from your general contractor stating the price that person will charge to build your house. Generally this contract should spell out the following:

1. The parties (who the contract is between). These are usually given as Owner and Contractor.

2. The time. This states when the contract work will begin and when it will end.

3. The sum. This states the total amount of the contract price. It should include the package price as well as all allowances (see item 5); it does not include land.

4. Progress payments. This corresponds to the draws on your construction loan from the lender (see chapter 10). The sum total of the progress payments should equal the contract price. Generally the lender inspects the site to determine that each phase of work has been satisfactorily completed, checks with you to see if you're satisfied, and then pays the contractor. If you're not satisfied with the workmanship or if you feel a certain phase is not complete, you may withhold a percentage of the progress payment as retainage. This is normal in the construction business and it assures everyone that the job will be completed as specified.

5. The allowances. These are budget *estimates* that you and the contractor agree will *vary* according to (a) factors encountered during the progress of the work on the site—such as the amount of fill needed to bring an area up to grade—and (b) miscellaneous factors—like what grade of carpeting you finally decide to buy or what appliances you will select. Allowances are best guesses based on changing circumstances and can often be supplied by your builder/dealer or manufacturer. They represent a figure plugged into the contract sum that *you* are responsible for. For example, if you establish a $4,000 figure for appliances and you can buy them for $3,000, you end up saving $1,000 on the contract price; if you establish a $1,000 figure for fill dirt and it costs $1,500, it will cost you—not the contractor—an additional $500.

6. The contract documents. This section states what documents are included in the contract, such as house plans and specifications, suggested builder specifications, and the schedule of finishes.

7. The Owner. This section states who owns the property where the house will be built and what the owner's responsibilities are, such as furnishing surveys and paying fees.

One-half of a new modular home is moved into place by a tractor.

It is then lifted onto the foundation by a crane.

One day later, a completed home awaits its siding.

8. The Contractor. This states the contractor's responsibilities, such as paying for all labor, using qualified persons, obtaining permits, and removing waste.

9. Settling disputes. This section states how disputes relating to the contract are to be settled. Let's hope that any dispute can be quickly settled by talking it out. If it can't, I would suggest that you consider controlling costs by pursuing arbitration and other techniques before resorting to a lawsuit. Most disputes can be equitably resolved by a professional arbitrator more quickly and cheaply than by going to court. Your lawyer, however, should look over your contract before you begin building to make sure that your interests are protected.

10. Correction of work and warranty. This section states the time period for which the contractor warrants all workmanship and materials (usually one year). You should have a separate warranty from the manufacturer for the package and all it includes.

11. Termination of contract. This section states under what conditions either the owner or the contractor may terminate the contract. For example, you may have gotten a builder who will not meet your expectations for quality work or who, for various reasons, cannot complete the job. If so, you will need a way to get rid of that builder and hire another.

12. Liability. There should be some clarification regarding both the owner's and the contractor's liability. Generally contractors provide insurance that protects them from claims under workmen's compensation acts and from claims for damages because of bodily injury. You, the owner, should have insurance that protects you from any contingent liability to others that may arise from operations under the contract. I would suggest that you shop around for a good insurance agent and let him or her advise you as to what you should carry. Confirm such advice with your lawyer.

13. Liens. Most states provide a mechanism whereby contractors and other subcontractors performing work on real property may claim a lien on property for the amount owed for their work. Because a contractor has an express contract with the owner, the contractor can create a lien against the property should the owner not pay the contractor. The same mechanism holds true for subcontractors. Because of this, your contract should have a clause that states that when the house is finished, the contractor will—at final payment—give you a complete release of all liens arising out of the contract.

14. Certificate of Occupancy. When the local building department is satisfied that your house has met all its requirements for a safe and habitable house, it will issue a certificate of occupancy to the contractor. The contractor will then turn this over to you in order to receive final payment. Usually this is the time when you, the new homeowner, make a final inspection of your new home to make sure that you're satisfied with all the

details. Any items that you see need correcting make up what's called a "punch list" that the contractor remedies before getting paid.

15. Change Orders. A change order is a written request from the owner to the contractor for a change in the construction plans or specifications of the house. Normally it involves relatively small items such as the addition of a window or skylight. But little things can add up and change orders help keep track of them and detail who pays for what. Both you and your contractor (and the subcontractors) should insist on them, as they help keep financial records accurate and eliminate financial disputes. All change orders requested by either owner or contractor should be in writing and signed by both parties before any changes are done. The price for the work should appear on the change order.

16. Additional Items. The information I've given here is not to be construed as a format for a legal contract; these are simply elements you will want to include. Some contractors have their own forms. The American Institute of Architects has a contract form that its members use with a contractor. Or you might want to draw one up yourself. Regardless of which approach you take, you should consult with an attorney when drawing up a contract. We live in a very litigious society, and your home is a big financial investment. The more comfortable you are with your contract, the better you'll sleep at night.

13

Doing It Yourself

One big advantage of factory-crafted homes is that they lend themselves more readily to a do-it-yourself approach than do site-built houses. There are modular manufacturers, for example, who will sell you a house that is 90 percent complete and will let *you* finish the remaining 10 percent. Many panelized, log, and dome products can be erected within a fairly short period by individuals who are relatively skilled with tools, able to read plans without too much difficulty, and have a fair knowledge of how things are supposed to go together.

Here, for instance, is a panelized dome that was put together by a group of friends in one day's time.

Using factory-crafted components, a dome shell can be quickly assembled by a small team.

But putting up a shell is one thing. Finishing it off is quite another.

Most states require, by law, that electrical, plumbing, and heating/air-conditioning work be done by licensed contractors and be subjected to both rough-in and finish inspections. In addition, I have found that such jobs as installing insulation and drywall are best left to those who have the special tools, supplies, and know-how. And anyone who has ever laid block or brick or placed a concrete slab knows that these jobs are hard, backbreaking work.

In addition, some lenders are very particular about risking their money on do-it-yourself homes. They prefer that the house be built by an experienced contractor. This way, if a default on the loan should occur, they feel that their investment will be protected. One friend, who heads up the lending department of a large regional bank, noted that he could recall only three recent instances where his bank gave construction loans to individuals who were not licensed contractors. Each customer, he said, was an experienced professional engineer who had taken a few courses in residential construction at the local community college. (He also admitted that all were employees of one of the bank's biggest customers.)

But this is not to put the kibosh on building your own home. The aforementioned lender works for one of the region's most conservative banks. In general, if you do intend to act as your own contractor, you can improve your chances of getting a loan by doing one of the following: (a) have enough equity so that the bank feels its risk position is greatly reduced or (b) do your homework so thoroughly that you have made all your estimates, have lined up all your major subcontractors (and have written copies of their quotes), have on hand all the manufacturer's plans and specifications, and give clear evidence that shows you know what you're doing and how you're going to do it. (See end of chapter 10 on making effective presentations.)

If you do want to use a contractor (or your lender insists that you do so) you may want to consider asking your contractor how you can pitch in and help. Some construction jobs, like painting and cleanup, don't require a high skill level and are not critical to the structural integrity of the house. Other jobs, like erecting the package, are more critical, but maybe you can provide some of the labor.

Steps in Building Your Home

If you are going to act as your own contractor you may want to consider either taking a course or two at the local community college or enrolling at one of the owner-builder schools listed in the Appendix on page 107. There is no great mystery to building a house, but there is a ton of details. It helps to have a good understanding of the general steps you will need to take, as well as the subcontractors who will help you take them.

Here is the sequence of steps you are likely to take in building your own home and a rough estimate of the average time for each step. Actual time will vary with weather conditions, different kinds of factory-crafted packages, and scheduling. Even if you use a contractor, you may find this list helpful in keeping track of your home's progress. The items with an asterisk apply to modular homes as well.

*1. Ordering the factory-crafted home or package: one hour.

*2. Getting permits: two to three hours.

*3. Staking the lot for clearing: one day.

*4. Clearing and rough grading the lot: one to two days.

*5. Ordering temporary utilities, portable toilet, getting insurance: three to four hours.

*6. Digging and pouring footings: one to two days.

*7. Soil treatment, foundation, waterproofing, and drains (if needed): one day to two weeks.

*8. Plumbing rough-in: one to two days.

*9. Slabs: one to two days.

*10. Package erection: one to four weeks. House erection if modular: one day.

*11. Chimney and fireplace: one week.

12. Roofing (including flashing): one to three days.

13. Plumbing, electrical, and heating/air-conditioning rough-ins (these trades can work at the same time): one to two weeks.

14. Insulation: 2 days.

15. Drywall: two to three weeks.

16. Interior trim: one to two weeks.

17. Hardwood flooring (installation): two to five days.

18. Painting and staining: two to three weeks.

19. Ceramic tile: one to two weeks.

20. Kitchen and bath cabinets and countertops: one to two weeks.

21. Plumbing, electrical, and heating/air-conditioning trim: two to four days.

22. Hardwood floor finish and/or carpeting: five days.

*23. Cleanup: two to three days.

*24. Driveway and walks: two to three days.

*25. Landscaping: one to three days.

*26. Final inspections: one to three days.

*27. Loan closing: one to two hours.

All the above steps need not be performed in this order. For instance, your brickmasons can be working on the chimney and fireplace while the mechanical rough-ins are taking place. Your hardwood floor installer and trim carpenters can work around each other. But you don't want sawdust flying while the painters are around. And you have to accept the fact that when the drywall finishers begin, you'll have to turn the house over to them while they're working (not much else can happen until they're done anyway).

Subcontractors

A subcontractor is an individual or firm that contracts to perform a specific part of the building process. Typical subcontractor trades include:

Surveyors survey your lot for grading and foundations.

Grading contractors remove trees and shrubs from your lot and provide a level building site.

Foundation contractors install footings and foundation walls.

Brickmasons install fireplaces, chimneys, and veneer.

Framing contractors erect a home's shell.

Roofing contractors install shingles, vents, flashing, and skylights.

Boxing and siding contractors install a home's exterior trim and siding.

Electrical contractors install all necessary wiring, outlets, switches, and lighting fixtures. They may also install some appliances.

Plumbing contractors install all necessary plumbing lines and plumbing fixtures. They may also install some appliances.

HVAC contractors install a home's heating, venting, and air-conditioning equipment.

Drywall contractors install and finish drywall.

Ceramic tile contractors install the tile surrounding a tub or shower and other tile flooring.

Painting contractors paint or stain your exterior and interior.

Concrete contractors install slabs, driveways, and walkways.

Landscape contractors plant trees, shrubs, grass, and sod.

Finding Good Subcontractors

Most mechanical trades (electrical, plumbing, HVAC) advertise in the yellow pages. Other subcontractors, however, are more difficult to find. One of the best trim carpenters I know just walked up to me on a job site one day, introduced himself, and left a card. Later I called him for help on a deck project, and we've worked together ever since. He doesn't advertise and doesn't have to. What's more, he has recommended several other subcontractors to me that he knows do good work.

If you're looking to find individuals like this, I'd suggest the following:

Ask real estate agents. They can often recommend good subs.

Drive around to different job sites and talk with workers. They can often help suggest good subs.

Check with suppliers of specific traders. I found three excellent drywall crews by asking the local supplier.

As you may have guessed by now, there is a network for contracting subs, but it is very casual and unstructured. Avoid the tacked up conglomeration of business cards at the local lumberyard or building-supply chain. The best subs are usually found by word of mouth. I like to apply what I call the appearance factor. A relatively neat and clean truck and personal appearance are usually good indicators of subs who apply the same standards to their work.

Subcontractor Bids

As with any other large financial expense, it is wise to shop around before committing yourself to buy. It sounds good to say "get three bids," but this isn't always possible or practical. Do get at least two, however, and make sure you compare "apples with apples." Do this by thoroughly communicating the scope of work and the specifications that apply to it. For example, if you want to get bids on installing your roof system you might say to each bidder: "I want a bid from you on installing 20 squares of architectural shingles, 100 feet of ridge vent, 65 feet of roll flashing, 30 feet of step flashing, and 2 skylights. I will supply all materials, and I want you to supply all labor and fasteners. I expect you to clean up the job site once you are finished. And I want a one-year written guarantee on your work." Check references, and I guarantee you will sleep better at night.

Subcontractors' Contracts

It used to be that a contract would consist of a person's word and his or her handshake. But nowadays unless you have previously worked with someone, you're better off getting your agreement in writing. To do this, you should consider a contract form that spells out:

(a) Who the parties are in this agreement: usually you and the subcontractor.

(b) A general description of the project: your house located at your address, city, state.

(c) The specific work desired: for example, "installing a concrete slab per attached specifications."

(d) Time frame for work: when the work will begin and when it will be completed.

(e) The price: how much you will pay the subcontractor for the work.

(f) Payment schedule. Sometimes a sub's work takes place over a few week's span. Mechanical trades (electrical, plumbing, heating and air-conditioning) are normally paid twice—after their rough-in work has been completed and inspected and after their trim work has been completed and inspected. Others, like framers, may want part of their money each week so they can pay their crew. Make sure you both agree to the terms. And make sure you factor in when you get *your* construction draw from the lender.

Some log home owners prefer to use wallboard and paint to finish off interior spaces.

(g) Construction documents. This section tells which construction documents (plans, specifications, or other) apply to a particular subcontractor's work. Sometimes subs may offer their own plans and/or specifications as amendments to yours.

(h) Other provisions. This is the section where you might want to detail what procedure(s) to follow with change orders. You should insist that any changes to the specific work be made only *after* both you and the subcontractor sign and date a change order spelling out the changes and the price for them.

As is the case with other legal documents, you will be well advised to have your attorney review your subcontractor agreement and change order forms *before* you implement them.

Scheduling Subcontractors

Once you have accepted bids from subcontractors, you can let them know when you would like them to start. Bear in mind that some construction tasks are critical and must follow in sequence (foundation, framing, roofing, insulation). Others can be done simultaneously after a critical task has been completed. For example, electrical, plumbing, and heating and air-conditioning rough-ins can be done at the same time after the shell has been completed and the house has been made weather tight. Often your subs can best tell you at what point they will want to start and will keep an eye on progress for you. Bear in mind that all subs are independent businesspeople who are looking to maximize their time and profits. If you have planned to use one and find that particular sub can't make your schedule, you will have to decide whether you want to wait or try to find another. Your decision should be based on how critical the task is to your home's completion and your likelihood of finding another qualified sub to do the work. If you've gone to a lot of trouble to find a good drywall contractor, for instance, and discover that person can't make it for another week, you may want to wait. On the other hand, if your painter can't come and finish painting the walls, according to your schedule, you may want to get another painter.

Working with Subcontractors

Once I have met with subs and feel that we understand the scope of work, I like to get out of the way and let the work get done. If I don't understand something, I ask questions. This gives subs a chance to impart their special knowledge, and I end up learning. If you have something special to be done, make sure you communicate it up front. If a sub suggests another way of doing it, listen first and then decide which way to go. Don't be afraid of insisting on what you want done if you have a specific reason for wanting it done that way. But if you don't, be flexible enough to go with what the sub suggests.

Inspecting the Work

Critical tasks—such as framing, electrical, plumbing, HVAC—are normally inspected by local authorities for quality and compliance with building codes. You can inspect less critical tasks—such as painting, drywall, and trim carpentry—by using common sense. If you are unhappy with the work, tell the subcontractor immediately. If the sub doesn't agree with your assessment, ask why. If you run into a situation where the work is obviously sloppy, your best bet is to fire the sub as soon as you can, pay what you both can agree is a fair amount, and get someone else to do the work. Be sure to write on your check that it is for a final payment. If the check is then cashed, you generally don't have to worry about having the sub coming back and say you owe more money. But if there's any questions, document the work with photographs and contact your attorney.

Most of the time, however, your experience should be pleasant. Subcontractors realize that reputations precede them, and they want to do a good job. And like others who take pride in their work, subcontractors like to be told when they've done a good job. It doesn't cost anything, and it makes for a more pleasant relationship.

Suppliers

As most factory-crafted home packages include all you'll need to erect a weather-tight shell, you'll be able to do without a great deal of lumber purchases. Other suppliers you will need to deal with as you build your house include:

1. Sand and gravel company: sand for your brickmasons, dirt for backfilling and landscaping, gravel for drives. Many subs supply these items, so you may not need this supplier.

2. Masonry supplier: for precast concrete block, face and cull (filler) brick, mortar mix. Sometimes you can get sand from this supplier as well.

3. Concrete supply company: Most concrete subcontractors will order concrete for you and figure it in their bid. But if you plan to do the work yourself, you will need this supplier.

4. Floor covering company: for carpeting, resilient, and hardwood floor material and installation.

5. Lighting and plumbing fixture suppliers: These can be specialty retailers or part of a major building supply retailer such as Lowes, Hechingers, and Home Depot.

6. Paint store: Your painter may give you a bid that includes materials, or you may want to specify and furnish them yourself.

7. Appliance store: If you don't already have these, you will want to shop around for the best brand and price for the appliances you'll need, such as refrigerator, microwave, oven, cooktop, garbage disposal, and dishwasher.

8. Tile company: for tub/shower surrounds, flooring, countertops, decorative exterior stone.

9. Cabinet supplier: for kitchen cabinets and countertops and bath vanities.

10. Hardware store: for all those extras like nails, glue, and tools.

   Some suppliers can meet many of your needs. For example, large chain stores, such as Lowes, HQ, Hechinger, and Home Depot, carry lumber, cabinets, lighting fixtures, paint, tools, and other building materials. But what they do have in breadth they often lack in depth, so you may also want to deal with specialty suppliers, like a plumbing supply house, that usually carry a greater variety of materials.

Appendix

Owner-Builder Schools

Recognizing the value of sweat-equity, a number of owner-builder schools have sprung up over the past decade or so. Their classrooms are often out on the job site where pupils learn by doing. Some offer intensive two-week courses, and their tuition includes room and board. Others have a different approach. What follows is a list of some owner-builder schools. This list is not an endorsement or recommendation and is included only as an aid, should you decide that the owner-builder route is for you.

Building Resources, 121 Tremont Street, Hartford, CT 06105. (203) 233–5165.

Colorado Owner Builder Center, PO Box 11158, Denver, CO 80211. (303) 433–8813.

Cornerstones at Nasson Institute, Box 850, Springvale, ME 04083. (207) 324–5340.

Heartwood Owner-Builder School, Johnson Road, Washington, MA 02135. (413) 623–6677.

Home Building Institute, 2424 North Cicero Avenue, Chicago, IL 60639. (312) 745–3910.

Michigan School of Home Building, 3135 South State, Ann Arbor, MI 48108. (313) 665–4321.

Owner Builder Center, 1250 Addison, Suite 209, Berkeley, CA 94702. (415) 848–6860.

Owner Builder Center of Sacramento, PO Box 739, 4777 Sunrise Boulevard, Fair Oaks, CA 95628. (916) 961–2453.

Owner Builder Centers, Inc., 3625 Point Elizabeth Drive, Chesapeake, VA 23321. (804) 483–4896.

Shelter Institute, 38 Center Street, Bath, ME 04530. (207) 442–7938.

Southface Homebuilding School, Box 5506, Atlanta, GA 30307. (404) 525–7657.

Urban Shelter, 717 Shelby Parkway, Louisville, KY 40203. (502) 635–7928.

Yestermorrow, Box 76A, Warren, VT 05674. (802) 496–5545.

Books

Here is a list of books that may prove useful. Appearance of a book on this list does not constitute an endorsement or recommendation but is intended solely as a guide for your information. For more complete references, visit your local library.

Alth, Max and Charlotte. *Be Your Own Contractor—The Affordable Way to Home Ownership,* Blue Ridge Summit, PA: Tab Books, 1984.

Browne, Dan. *The Housebuilding Book,* New York: McGraw Hill Book Company, 1974.

Hasenau, J. James. *Build Your Own Home—A Guide for Subcontracting the Easy Way,* Northville, MI: Holland House Press, 1973.

McGuerty, Dave, and Kent Lester. *The Complete Guide to Contracting Your Home,* White Hall, VA: Betterway Publications, Inc., 1986.

Roskind, Robert. *Before You Build—A Preconstruction Guide,* Berkeley: Ten Speed Press, 1981.

Industry Trade Associations

Manufactured Housing Institute, 1745 Jefferson David Highway, Suite 511, Arlington, VA 22202. (703) 979-6620.

Mortgage Bankers Association, 1125 15th Street, N.W., Washington, D.C. 20005. (202) 861-6500.

National Assocation of Home Builders (Modular Building Systems Council, Panelized Building Systems Council, North American Log Homes Council, National Domes Council), 15th and M Streets, N.W., Washington, D.C. 20005. (202) 822-0576.

National Association of Real Estate Brokers, 1629 K Street, N.W., Washington, D.C. 20006. (202) 785-1244.

Index

Carpentry
  framing, 83, 99, 102-4
  trim, 85, 99, 100, 101, 104
Ceramic tile, 18, 20, 85, 99, 105
Certificate of Occupancy, 95
Chamber of Commerce, 92
Change Orders, 96
Charlotte Observer, 30
Chimney and fireplace, 84, 99, 100. *See also* Masonry.
Cleanup and disposal, 86, 98, 99
Closed wall, 40, 41, 43. *See also* Panelized homes.
Colorado, 34
Computer-aided design (CAD), 2
Concrete
  driveway, walks, patio, 18, 22, 86, 99
  slabs, 82, 86, 98, 99, 102
  suppliers, 104
Construction (temporary) loan, 15, 19, 22, 53, 75, 78, 98
Contracts
  as documents, 93
  fees for, 90
  general conditions of, 93-96
Costs, 79-87
  closing, 81
  estimating of, 79-87
  interest, 19, 81
  modular, 19
  package, 82
  plans, 14, 81
  shipping, 14
Crane, 19, 61, 62, 83

D
Decks, 18, 22, 84
Deed, 78
Department of Veteran Affairs (VA), 22, 32, 34, 53
Dome houses, 5, 12, 13, 59-63, 69, 74, 77, 79, 80, 97

W

Warranties, 15, 95
  Home Owners Warranty Program, 15
Winter, Steven, 5

Y

Yellow pages, 34, 92, 101
Yurts, 60

Z

Zoning, 30, 34

About the Author

Paul Sedan has written extensively about architecture and homebuilding for a variety of local, regional, and national publications. As a licensed general contractor he also heads up his own firm, which builds, among other things, factory-crafted homes. He has lived and worked in Detroit, New York, and Boston and currently resides with his wife, daughter, and family cat in Charlotte, North Carolina.